# System Safety Primer

Clifton A. Ericson II

# CONTENTS

1   Safety   1

2   System Safety   7

3   System Safety Program (SSP)   23

4   Core System Safety Process   29

5   Systems and Systems Laws   45

6   The Hazard-Mishap Relationship   57

7   Hazard Theory   61

8   Risk Theory   71

9   Hazard Identification   77

10   System Safety Tools   87

11   System Safety Values, Axioms and Principles   95

12   Design Safety Methods   103

13   Common Mistakes in System Safety   115

14   Software Safety   117

15   System Safety Validation   123

A   Example Hazard Analysis   129

B   Example Risk Rating Methodology   135

  Author Biography   139

# FOREWORD

We assume every day of our lives that we are safe, absent situations that seem very apparent (i.e., military conflict, crime, etc.). We expect that systems we use are safe. However, it is frightening to realize designers and manufacturers of systems rarely consider system safety relative to their designs. It is further frightening that most engineers, designers, manufacturers vaguely understand the concepts of system safety that have, in some form or another, existed for decades.

Engineers, designers and manufacturers of systems have been taught to focus their efforts on making sure the system functions according to plan. Very few engineering schools focus their efforts on system safety. Therefore, it is not surprising engineers have little understanding of system safety and the importance thereof. Most engineers believe that designing a system to meet a consensus standard of industry is somehow satisfactory for assuring that a system is safe.

Chief engineers for manufacturing industries need to read and embrace this primer. Too many times, industry continues to design a system the same way, disregarding the concepts of system safety. Only when there is a failure which results in serious injury and/or death or significant property damage does industry attempt a retrospective analysis of what happened, mostly, undertaking an analysis to focus on human error rather than the system itself. Had system safety been incorporated in a proactive manner, in all probability the hazard, risk and danger which resulted in the failure would have been discovered and addressed long before the failure had an opportunity to occur.

This primer needs to be read by the insurance industry and its high-level risk managers. System safety reduces losses for which insurance carriers would be required to pay. If insurers would require its insured's to undertake system safety, catastrophic payments to those injured for loss of property would be significantly reduced. Risk managers within the insurance industry would have a new understanding of their roles, not to manage risk, (i.e., probability of an undesired occurrence) but rather the elimination of hazards (i.e., unsafe physical conditions that can result in potential injury or property damage). These managers would also understand that there is never a true "accident." That is, there is always a reason something happens, which, in all probability, system safety methodologies would identify long before the incident. These managers would also understand that the word "accident" is an irresponsible way out for avoiding accountability and responsibility.

Professionals within the legal system need to read and understand this primer. With all due respect to numerous judicial opinions, most courts do not understand system safety and have an incorrect analysis as to how systems can be made safe. Too many times, courts misapply the terms hazard, risk, danger and fail to understand the system safety hierarchy for the design, manufacture, and distribution of products. This leads to a further misunderstanding by legal professionals in the analysis of legal matters.

In my 25 years of legal work in the courtroom, the majority of which has been on behalf of persons severely injured and/or killed, I have read many articles regarding system safety and searched literature for concise authoritative sources regarding the topic. Such sources have been difficult to find until now. This primer written by Clifton is an excellent source for those who are truly committed to system safety and, thus, a safer society.

I always remind myself that a mind is like a parachute: it works best when open. This primer, if read with an open mind, will enlighten those who are truly interested in a safer society.

J. Kevin King
Cline, King & King, P.C.
Columbus, Indiana
August 2011
www.lawdogs.org

# PREFACE

It's a fact of life – we are continuously exposed to hazards and risks. We interface with systems on a daily basis, and systems expose us to many different types of potential mishaps. Most systems involve hazardous components and/or safety-related functions, which means that these systems will naturally have inherent hazards as part of their design. These hazards must be eliminated or controlled if mishaps are to be prevented.

In order to eliminate or mitigate a hazard, the hazard must first be identified. This can be done in one of two ways: a) by reactively identifying the hazard after a mishap or b) by proactively identifying the hazard before a mishap occurs through hazard analysis and the system safety process. Reacting after a mishap is contrary to the principles of system safety.

System safety is an engineering discipline that is applied during the design and development of a product or system to identify and eliminate/mitigate hazards, and thereby eliminate/reduce the risk of potential mishaps and accidents. System safety is ultimately about saving lives. It is a proven technique that is currently applied to a diversity of systems, such as commercial aircraft, military aircraft, ships, trains, automobiles, nuclear power plants, weapon systems, chemical processing plants, mining, software, medical devices. The lack of system safety costs millions of dollars in damages and loss of lives every year due to preventable mishaps. It is my greatest hope that the readers of this book can use the material contained herein to better understand and apply the system safety technique and thereby develop safe products and systems.

The practice of system safety requires detailed knowledge and understanding of the tools, techniques and processes involved in the system safety discipline. I have written this book for engineers, analysts and managers who are confronted with the responsibility of developing safe systems and products. The book provides a brief introduction to the system safety process; it is an overview as opposed to a detailed instruction. This book is intended for persons from various industries who are interested in making safe products and systems. It should be very useful to those individuals new to the system safety discipline with a desire to understand the basic methodology. It is also intended as a refresher for system safety practitioners who already apply the system safety process in their daily job.

Many individuals think safety is easily achieved by just applying common sense, standards and good design practices, but it turns out that achieving true safety requires more. Systems are becoming too large, too complex and contain too many hazardous aspects to be left to designers inexperienced and untrained in the specialty of system safety. The general concept of system safety is incredibly simple; however, the implementation is more intricate because it requires time, resources, effort and special expertise. The system safety process has many benefits and is the most cost-effective approach in the long run for achieving a safe system design. Safety must be bought upfront and be designed into the system or it will be paid for after the system is developed by paying for mishaps and design changes to add safety back in. Safety is no accident.

Clif Ericson
Fredericksburg, VA
July 2011
www.risk-logic.com

# DEDICATION

This book is dedicated to the loving memory of Jasper Ericson. It is also dedicated to the many system safety professionals who strive to make safe systems in order to save lives and make the world a safer place.

# 1   SAFETY

1.1 Introduction to Safety

We live in a world where we are constantly exposed to hazards and risks. This is significant because hazards ultimately lead to mishaps. Mishaps take lives, cost money and have significant repercussions and collateral effects. Hazards and risk exist at all levels of life; mishaps have become commonplace events that are often accepted as normal events. As technology advances, hazards, risks and mishaps also seem to advance and increase. It sometimes seems like a hopeless spiral into the acceptance of random mishaps and deaths.

On the positive side, hazards, risks and mishaps can be intentionally prevented and controlled. Humans have a natural instinct or predilection for self-preservation, which could be called the need for safety. The endeavor for safety has been around as long as mankind. For example, the ancient Code of Hammurabi had 282 dictums or laws, one of which stated, "If a house falls in and injures the occupants, then the builder shall be put to death". Throughout history mankind has sought to live and work in safe environments and situations through specialized engineering means. However, this is not always an easy task, as demonstrated by the Titanic, the ship designed to be the safest in history. But the task is not impossible.

The key to understanding safety is to recognize that a mishap (aka accident) is not just an unexplainable random event that just occurs out of nowhere. There is always a set of causal factors involved, and these causal factors are known as hazards, which predict and control potential mishaps. The goal to developing safe systems is to eliminate or minimize system

1

hazards so they do not result in mishaps. The most successful way to achieve this objective is to apply the engineering discipline known as **system safety**. The system safety process forces system developers to intentionally design and build safety into a system (or product). A known and acceptable level of safety can be achieved when the system safety process is consciously and unconditionally applied.

Thinking of a mishap as a chance event without cause gives one the sense that mishaps involve an element of destiny and futility. System safety, on the other hand, is built upon the premise that mishaps are not just chance events; instead they are seen as deterministic, predictable and controllable events (in the guise of hazards). System safety demonstrates that we do have control over potential mishaps in the systems we develop and operate. We are not destined to face unknown mishaps, unless we allow it to be so (by not performing adequate system safety). In the safety sense, mishaps are pre-planned events, in that they are actually created through poor design and/or inadequate design foresight.

There is more to safety and building safe systems than just common sense, design safety standards and standardized design processes. Achieving safety requires a specialized process – system safety. Systems are becoming too large, too complex and contain too many hazardous aspects to be left to designers inexperienced and untrained in the specialty of system safety. The general concept of system safety is incredibly simple; however, the implementation is more intricate because it requires special expertise and training in system safety engineering.

Safety is an attribute of a system indicating the level of danger the system presents. For example, an aircraft that is experiencing a large number of mishaps demonstrates a poor safety attribute, which can be measured by the number of hazards associated with the design. Safety is also the act of eliminating or reducing the danger presented by a system, which is achieved by eliminating and controlling hazards. The safety attribute of a system is variable; when hazards are eliminated or mitigated the safety attribute is improved.

The system safety process has many benefits and is the most cost effective approach in the long run for achieving a safe system design. Safety must be bought upfront and be designed into the system or it will be paid for after the system is developed by paying for mishaps and design changes to add safety back in. We interface with systems on a daily basis, and systems expose us to many different types of potential mishaps. Most systems involve the necessary use of hazardous components and safety-related functions, which means that these systems will have inherent hazards as part of their design. In order to prevent mishaps, these hazards must be identified and then eliminated or mitigated. This can be done in one of two ways: a) by reactively identifying the hazard after a mishap or b) by proactively identifying the hazard before a mishap occurs through hazard analysis and employing the system safety process. Reacting after a mishap is contrary to the principles of system safety.

The bottom line is that mishaps can be eliminated, or their risk reduced, when the proper system safety processes are conscientiously applied. Safety must be earned through the rigorous application of the system safety process. It is not free, but it is cheaper than the cost of mishaps. System safety is a formal and systematic engineering and management process for proactively making a system safe.

## 1.2 Key Safety Terms

There is often confusion, misunderstanding and misuse of safety related terms. Some terms are colloquial in usage and can be easily misused or misinterpreted. In addition, the colloquial definitions are somewhat weak when used in engineering applications dealing with safety and hazard elimination/mitigation. For purposes of clarity and common understanding, some of the basic safety-related terms are defined here as they are used in the system safety discipline. The definitions of these terms will be expanded as they are explained in more detail in the following chapters.

### Safe

*Safe* is typically defined as free from danger or the risk of harm; secure from danger or loss. Safe is a state of being that is secure from the possibility of death, injury or loss. A person is considered safe when there is little danger of harm threatening them. A system (product) is considered safe when it presents low mishap risk (to users, bystanders, environment, etc.). Safe can be regarded as a state... a state of low mishap risk (i.e., low danger); a state where the threat of harm or danger is non-existent or minimal. How safe something is varies and is measurable; every situation and scenario has a different level of safeness, which depends upon the particular variables involved. Relative safeness is calculated by the metric of *risk*. Risk is the estimated value of a potential future event, based on the event's gain or loss and the event's likelihood of occurrence. Thus, how safe something is becomes a function of the amount of risk involved.

Safety

*Safety* is freedom from those conditions that can cause death, injury, occupational illness, damage to or loss of equipment or property, or damage to the environment (MIL-STD-882D). Since 100% freedom from hazards and risk is not possible, safety is more effectively defined as *"freedom from unacceptable mishap risk"*. Safety is the *condition* of being protected against physical harm or loss, with some level of risk involved, but the level of risk is considered acceptable. Establishing that safe *level* of risk is part of the system safety process. Safety is an *attribute* implying the "safeness level" of something (i.e., potential mishap risk). Safety has also become a term used to describe the *act* of making something safe; it is the act or steps taken to control the safe attribute of something (i.e., eliminating or mitigating hazards). The term safety is often used in various casual manners, which can sometimes be confusing. For example, "the designers are working on aircraft safety" implies the designers are establishing the conditions for a safe state in the aircraft design. Another example, "aircraft safety is developing a redundant design" implies an organizational branch of safety, "aircraft safety" that is endeavoring to develop safe system conditions. A "safety device" is a special device or mechanism used to specifically create safe conditions by mitigating or controlling the risk of an identified hazard.

Danger

*Danger* is the unreasonable or unacceptable combination of a hazard and risk[1]. Danger is the condition of being susceptible (or exposed) to harm or injury. Danger is present in the form of an existing hazard which presents a potential undesirable outcome and probability of occurrence (risk). Danger characterizes the relative exposure to a hazard and therefore to a potential mishap. The amount of danger presented by a hazard depends upon the amount of risk imparted by the hazard, which is a function of the parameters involved. Danger and risk can be modified by modifying the existing parameters and conditions.

---

[1] H. M. Philo, et al, *Lawyers Desk Reference*, 9th edition, 2001, volume 2, chapter 5, page 4.

Absolute Safety

*Absolute safety* is not clearly defined anywhere, but it would seem to mean complete 100% safeness or zero risk. This would imply that no hazards exist. Given this understanding, absolute safety does not seem realistic. In some situations absolute safety may be achievable, but in most situations it is typically not possible. This is why system safety exists, to reduce and control the safety risk in those situations where absolute safety is not possible. Acceptable safety risk is not beyond human control, but absolute safety is not realistic. Safety is really not an absolute, but instead is a relative value.

Mishap

A *mishap* is an unplanned event or series of events resulting in death, injury, occupational illness, damage to or loss of equipment or property, or damage to the environment (MIL-STD-882). It should be noted that in system safety the terms *mishap* and *accident* are synonymous. Mishaps are often assumed to be stochastic events, i.e., random, haphazard, unpredictable. However, a mishap is not just a random unplanned event with an unpredictable free will. Mishaps are not events without apparent reason; they are the direct result of specific causal factors which are defined by hazards.

Accident

An *accident* is an unplanned event, or events, that culminate in death, injury, damage, harm and/or loss. Accidents are the result of specific causal factors, which are defined by hazards. There is a direct relationship between hazards and accidents. For example, the crash of an aircraft is an accident event, with death of the occupants and loss of the aircraft the resulting outcome. This accident could be the result of a hazard such as insufficient fuel during flight to allow for a safe landing. An accident is not just an unexplainable random event; there is always a set of causal factors that explain the reason it happened. The term "accident" should not be used as a way of disregarding responsibility and accountability for an accident event. Accident prevention is not beyond human control.

Incident

An *incident* is the occurrence of an unexpected event, generally undesired, where the outcome does not result in damage, injury, death or loss. An incident is considered to be a near miss to a mishap, which would have more serious consequential outcome. An incident is similar to a mishap in that it is the outcome resulting from an actualized hazard; however, in this case the hazard is only partially actualized thus precluding serious loss or damage. If complete hazard actualization had taken place a mishap would have resulted.

## Hazard

A *hazard* is any real or potential condition that can cause injury, illness, or death to personnel; damage to or loss of a system, equipment or property; or damage to the environment (MIL-STD-882). A hazard is a source of danger; it is a potential condition that can result in a mishap. The hazard condition always comprises three necessary components: a hazard source, an initiating mechanism and an undesired outcome. The hazard condition is dormant, but it transforms into a mishap when the inactive hazard state components are activated. It is necessary to identify and understand the hazard components in order to determine the risk involved and to eliminate or mitigate the hazard.

## Risk

*Risk* is a metric quantifying the amount of danger presented by a hazard; it is a composite of hazard likelihood and hazard severity. Risk characterizes the uncertainty and threat presented by a hazard. Risk can be changed when the factors comprising a hazard are modified (by design means).

## System Safety

*System safety* is as an engineering methodology employed to intentionally design-in safety into a product or system through the identification and elimination/mitigation of hazards. System safety is a well-defined systematic approach to safety that applies a *systems* oriented approach in order to intentionally understand, modify and control the safety attribute of a system from a total systems perspective. System safety proactively designs safety into the system and establishes a safety case verifying the design safety tasks performed. See chapter 2 for additional detail.

# 2    SYSTEM SAFETY

2.1 System Safety Introduction

Forensic engineering is the detailed investigation of a mishap after it has occurred, performed to determine the specific causes for the mishap in order that corrective action can be applied to prevent reoccurrences. System safety, on the other hand, is a form of preemptive forensic engineering, whereby potential mishaps are identified, evaluated and controlled before they occur. Potential mishaps and their causal factors are anticipated and identified during the design stage, and then design safety features are incorporated into the design to control the occurrence of the potential mishaps – safety is intentionally designed-in and mishaps are effectively designed-out. This proactive approach to safety involves hazard analysis, risk assessment, risk mitigation through design and testing to verify the design results. Potential mishaps are recognized and identified by the hazards which ultimately cause them. System safety is a proactive approach to affecting the future (i.e., preventing mishaps before they occur) by identifying hazards and then eliminating or controlling the risk they present.

On the surface, creating a safe product seems a relatively simple and easy task … just apply design standards and a design process and it happens automatically. However, there are many intricate factors involved in a product/system design that necessitate a formal safety process in order to affect and control the safety attribute (i.e., risk) of a system design. Lessons learned through the years have shown that in order to enhance the safety attribute of a product or system, a total *systems* approach is required.

7

Individuals unfamiliar with system safety typically think that hazards result only from failures, errors and extreme environmental conditions. One of the enigmas of safety is that hazards can exist with or without the occurrence of failures or errors. Two perfectly working components, when combined in a system, can actually result in a system hazard. Also, a failure in two different components may not be hazardous if the components are used alone, but when the two components interact in the system, their combined failure modes can be hazardous. Sneak system paths and unsuspected common cause events can wipe out intended safety features. These are reasons why perfect design requirements or generic standards will never guarantee a safe system design, because they cannot eliminate these unseen hazardous design conditions. Thus, the only way to ensure a safe design is to identify and mitigate these unseen design conditions (i.e., hazards) that exist within the actual system design configuration. System safety is evidence-based; evidence of safety is required, from hazard identification to hazard mitigation to mitigation verification testing; the system is guilty until proven innocent with a safety case containing the evidence.

## 2.2 System Safety Background

System safety is an engineering discipline for developing safe systems and products, where safety is intentionally designed into the system or product. It involves the planned application of management and engineering principles, criteria, and techniques for the purpose of developing a system that presents acceptable mishap risk. System safety applies to all phases of the system life cycle and covers all system aspects, such as hardware, firmware, software, human operators and procedures. System safety is the process for eliminating or reducing potential mishaps through a process of hazard identification, safety risk assessment and safety risk management. System safety is holistic and interdisciplinary in nature.

System safety is a well thought out process that is planned, proactive, prudent and preventive in nature. The primary objective of system safety is to avert mishaps by ensuring that safety is intentionally designed into a product or system. This is accomplished by eliminating or controlling hazards that reside within the system architecture. System safety is officially defined in U.S. Department of Defense (DoD) document MIL-STD-882D, dated 10 February 2000, as:

"The application of engineering and management principles, criteria, and techniques to achieve acceptable mishap risk, within the constraints of operational effectiveness and suitability, time, and cost, throughout all phases of the system life cycle".

System safety engineering was defined in U.S. DoD document MIL-S-38130 (the predecessor to MIL-STD-882) as:

"An element of Systems Management involving the application of scientific and engineering principles for the timely identification of hazards and initiation of those actions necessary to prevent or control hazards within the system. It draws upon professional knowledge and specialized skills in the mathematical, physical, and related scientific disciplines, together with the principles and methods of engineering design and analysis to specify, predict, and evaluate the safety of the system".

MIL-STD-882 and its predecessor MIL-S-38130 are the genesis of system safety. The U.S. military, along with U.S. aerospace companies, saw the need for a holistic and proactive "systems" approach for the design, development, test and manufacture of "safe" systems. Working together, these two groups developed the system safety methodology and discipline. MIL-S-38130 was originally released on 30 September 1963 and replaced by MIL-STD-882 on 15 July 1969. System safety was actually documented as a process prior to any formal documentation of the systems engineering discipline. System safety as a formal discipline was originally developed and promulgated by the military-industrial complex to prevent aircraft and missile mishaps that were costing lives, dollars and equipment loss. As the effectiveness of the discipline was observed by other industries, it was adopted and applied to these industries and technology fields, such as commercial aircraft, nuclear power, chemical processing, rail transportation, the FAA and NASA, just to name a few.

2.3 System Safety

System safety is a specialized engineering discipline for developing safe systems, products, processes, procedures and operations. Safety is characterized by a metric called "risk", which indicates the danger level (susceptibility to harm) presented by a system hazard. A system is considered "safe" when the inherent risk it presents is known and considered acceptable (i.e., low risk). System safety involves applying proven engineering and management safety principles to intentionally design-in and build-in safety from the start of system development, as opposed to trying to add it in at a later time (after encountering incidents and mishaps). System safety anticipates the undesired outcomes that can result from failures, errors and design flaws and establishes design safety measures to counter these potentially hazardous situations.

System safety has many aspects and purposes; however, overarching goals of system safety consist of the following:

- To proactively identify and eliminate/mitigate hazards
- To save lives and preclude monetary losses by preventing system mishaps
- To protect the system and its users, the public and the environment from mishaps
- To design and develop systems presenting minimal mishap risk
- To intentionally design safety into the overall system fabric
- To reduce costs by building-in safety from the start, rather than adding it later

System safety is effectively a Design-for-Safety (DFS) process, discipline and culture. DFS means that the design process utilizes the system safety process to intentionally design-in safety. This process anticipates potential safety problems (i.e., hazards) and eliminates them or reduces the risk they present. Safety risk is calculated from the identified hazards, and risk is eliminated or reduced by eliminating or mitigating the appropriate hazard causal factors. System safety, by necessity, considers function, criticality, risk, performance and cost parameters of the system. Risk mitigation is achieved through a combination of design mechanisms, design features, warning devices, safety procedures and safety training to counter the effect of hazard causal factors.

System safety involves a systems approach, which accounts for the distinctive name. System safety is the art and science of looking at all aspects and characteristics of a system as an integrated whole, rather than looking at individual components in isolation from the system. System safety is a holistic approach that considers the subject as an integrated sum-of-the-parts combination, rather than a piecemeal approach of looking at separate individual and solitary pieces of the system.

System safety is often not fully appreciated for the contribution it can provide to creating safe systems that present minimal chance of deaths and serious injuries. System safety applies a planned and disciplined methodology for purposely designing safety into a system. A system can only be made safe when the system safety methodology is consistently and properly applied. Safety is more than eliminating hardware failure modes; it involves designing the safe system interaction of hardware, software, humans, procedures and the environment, under all normal and adverse failure conditions. Safety must consider the entirety of the problem, not just a portion of the problem, i.e., a systems perspective is required for full safety coverage. System safety anticipates potential problems and either eliminates them, or reduces their risk potential, through the use of design safety mechanisms applied according to a safety order of precedence.

The system safety concept is characterized by the following traits, which help describe the inherent aspects of the system safety process:

- Safety Focused – The primary goal is to save lives; acceptable safety is not sacrificed for cost and schedule.
- Proactive – The process identifies and mitigates safety issues from the start of product design rather than trying to eliminate them after a mishap occurs.
- Preventive – Hazards are anticipated and mitigated; the safety quality is intentionally designed into the product to prevent potential mishaps.
- Hazard Based – The process concentrates on hazard identification and mitigation because hazards are the key to potential mishaps and risk.
- Risk Oriented – Safety is measured by the metric of risk, and a risk management approach is utilized to mitigate and control potential hazard-mishap risk.
- System Based – The process focuses on the system as a whole, rather than just individual parts of the system, because of the many complex interactions involved between hardware, software and humans.
- Lifecycle Oriented – The process focuses on the entire product/system lifecycle for optimum product/system risk.
- Process Based – The system safety process follows a defined, rigorous, structured and disciplined best practice methodology that has been proven over the years.

- Integrity Based – Engineering integrity and ethics are applied to produce a product/system that is safe, taking precedence over cost, schedule and profit.
- Evidence Focused – It must be shown through evidence that an attempt has been made to identify all hazards and that they have all been appropriately eliminated/mitigated and formally accepted.

System safety influences and controls the safety attribute of a system, such that the attribute is often referred to as an *emergent property* of a system. This emergent attribute or property can vary according to the design safety feature applied to the system design. The metric for this emergent property is hazard/mishap risk, which can be modified by design methods.

There are many myths and false ideas about safety and system safety. System safety is a formal engineering process that can only be effectively applied by skilled system safety engineers or analysts applying the process described herein.

System safety *IS*:

- The application of a rigorous, defined and documented process
- The focused identification and elimination/mitigation of hazards
- The proactive prevention of mishaps
- A hazard and risk based approach based on evidence and traceability
- An integrated systems-focused approach
- A team approach involving a system safety expert and technical area experts

System safety *IS NOT*:

- A mere review of incident/mishap statistics
- Routine compliance with codes/standards (prescriptive safety)
- Applying reliability or quality methods
- Routine application of engineering design practices
- Letting insurance pay for lack of safety effort
- Everyone's responsibility (where there is no rigor, responsibility or accountability)

System safety is not prescriptive safety, which is routine compliance with standards and regulations. System safety does utilize known safety requirements and guidelines for products and systems; however, it has been proven that compliance-based safety alone is insufficient for complex systems because compliance requirements do not cover subtle hazards created by system complexities. Also, new technologies may spawn new hazards not covered by old standards. To achieve a known level of safety risk all hazards must be identified, assessed for risk and mitigated to a known level of risk.

System safety is not occupational health safety. Occupational safety is an important safety discipline that deals with safety during the performance of a job or work activity. Occupational safety can apply system safety as part of its methodology, but system safety is much bigger and broader in scope. System safety attempts to design safety into a product or system before it is put into operational usage; operations safety is actually impacted during system design, long before the system becomes operational. System safety is not the same as reliability and cannot be supplanted or achieved strictly by reliability. Making a system reliable does not necessarily make it safe, and making a system safe does not necessarily make it reliable. Typically, system safety and reliability work well together; however, there are many situations where enhanced reliability actually degrades safety, and vice versa.

2.4 Why Is System Safety Needed?

We live in a world surrounded by hazards and potential mishap risk; they are a reality of daily life. One of the major reasons for hazards is the ubiquitous system; many hazards are the byproduct of man-made systems, and we live in a world of systems and systems-of-systems. Systems are intended to improve our way of life, yet they also contain the inherent capability to spawn many different hazards that present us with mishap risk. It's not that systems are intrinsically bad; it's that systems can go astray, and when they go astray they typically result in mishaps. System safety is about determining how systems can go bad and implementing design safety mitigations to eliminate, correct or work around safety imperfections in the system.

Murphy's Law states that "if anything can go wrong, it will". This truism illustrates that the unexpected and undesired must be anticipated and controlled in order to prevent mishaps, and this can only be achieved through the system safety process. Hazards and risk often cannot be eliminated; however, hazards and risk can be anticipated and mitigated via safety design features, thereby preventing or reducing the likelihood of mishaps. If system safety is not applied, accidents and loss of life will not be prevented. System users are typically not aware of the actual risk they are exposed to, and without system safety this risk may be much higher than the users realize.

Systems fail and become unsafe for various reasons. Quite often it is not possible to eliminate these reasons, but they can be controlled when they are known and understood. Potential mishaps exist as hazards in system designs. Hazards are inadvertently designed into the systems we design, build and operate. In order to make a system safe, we must first understand the nature of hazards in general and then identify hazards within a particular system. Hazards are predictable, and if they can be predicted they can also be eliminated or controlled, thereby preventing mishaps.

Systems seem to have both a bright side and a dark side. The bright side is when the system works as intended and performs its intended function without a glitch. The dark side is when the system encounters hardware failures, software errors, human errors and/or sneak circuit paths that lead to anything from a minor incident to a major mishap event. The following are examples of the dark side of a system, which demonstrate different types and levels of safety vulnerability:

- A component in a toaster fails, causing the toaster to overheat, and the thermal electrical shutoff fails, allowing the toast to burn, resulting in flames catching low hanging cabinets on fire, which in turn results in the house burning down.
- A dual-engine aircraft has an operator-controlled switch for each engine that activates fuel cutoff and fire extinguishment in case of an engine fire. If the engine switches are erroneously cross-wired during manufacture or maintenance, the operational engine will be erroneously shut down while the other engine burns during an engine emergency.
- A missile system has several safety interlocks that must be intentionally closed by the operator in order to launch the missile; however, if all of the interlocks fail in the right mode and sequence, the system will launch the missile by itself.
- Three computers controlling a fly-by-wire aircraft all fail simultaneously due to a common cause failure (such as exposure to excessive RF radiation), resulting in the pilot being unable to correctly maneuver the flight control surfaces and land the aircraft.
- A surgeon erroneously operates on the wrong knee of a patient due to the lack of safety procedures, checklists, training and inspections in the surgical process.
- A software error combined with a unique computer hardware glitch in the throttle control system of an automobile result in uncontrolled full throttle being applied to the auto, without operator input or control.
- A crane is raised into overhead high voltage lines that are covered by trees and not readily visible; electrocution of the operator results

because the crane has no protective safety devices to warn the operator and no insulated covering to prevent an electrical path to ground.

Hazards and mishap risk will always be with us because of some natural system laws that exist. These laws include the following items that can lead to hazards:

- Eventually everything physical fails or wears out.
- Human error will always occur.
- Design errors occur.
- Hazardous system elements are used within systems for needed system functions, and these hazard sources precipitate hazards (e.g., fuel spawns many different hazards).

Since many systems and activities involve hazard sources that cannot be eliminated, zero mishap risk is often not possible. Therefore, the application of system safety becomes a necessity in order to reduce the likelihood of mishaps, thereby avoiding deaths, injuries, losses and lawsuits and potentially reducing insurance costs. Safety must be designed intentionally and intelligently into the system design or system fabric; it can't be left to chance or forced in after the system is built. If the hazards in a system are not known, understood and controlled, the potential mishap risk may be unacceptable, with the result being the occurrence of many unanticipated mishaps.

2.5 The System Safety Concept

System safety is a process for conducting the planned application of management and engineering principles, criteria, and techniques for the purpose of developing a safe system. Acceptably safe systems can be developed through the following three basic steps:

1) First, explicitly define safety and system safety as core values of the company or organization and stand behind this value. Establish a system safety culture and use a technically qualified system safety staff. Management must ensure there is proper funding and resources for the SSP.

2) Second, prevent any initial unnecessary hazards in the system design. This is accomplished by facilitating the system safety analysis required to identify hazards and successfully design-out those hazards that can be effectively removed. It is important to sensitize design engineers to be attentive to system hazards while creating the design,

so they may minimize the number of hazards initially residing in the system.

3) Third, manage residual hazards that cannot be eliminated and therefore remain in the system design (for valid reasons). This is achieved by mitigating the residual risk of hazards that cannot be eliminated and assuring the proper level of management awareness and acceptance of these risks.

The objective of system safety is to develop a system that is considered safe and provides acceptable minimum mishap risk. The basic system safety philosophy for achieving this goal is to confront hazards, mishaps and risk at three different levels of safety defense. The safety levels of defense that are engineered into the system design include:

- Design the system to operate safely under normal operating conditions.
- Design the system to safely tolerate abnormal operations caused by faults and errors.
- Design the system to provide survival protection from foreseeable mishaps.

System safety is more than eliminating hardware failure modes; it involves designing the safe interactions of hardware, software, humans and the environment under all failure and adverse conditions, as well as normal operating, testing, handling, or maintenance conditions. System safety involves anticipating potential failures and human errors and designing to safely counter the threats they present.

The objective of system safety is to make it impossible or very difficult for mishaps to occur, and to minimize the consequences should they occur. Mishaps are more likely to occur in certain types of systems because of failures in the design of the system. Perrow[2] characterizes systems according to two characteristics: complexity and coupling. Systems that are more complex and more tightly coupled tend to be more prone to mishaps (i.e., hazards exist within the design). In complex systems, one component can interact with multiple other components, in both intended and unintended ways. Complex systems are characterized by redundancy, specialization, interdependencies, multiple feedback loops and information flow. Tight coupling means there is little or no buffer between two components; processes are more time dependent and event sequences are more rigidly

---

[2] Perrow, Charles, *Normal Accidents*, New York, Basic Books, 1984.

fixed. Loosely coupled systems can tolerate timing delays and data latency, as well as changes in event order. As Reason[3] summarizes, complex and tightly coupled systems "can spring nasty surprises". It is the task of system safety to discover these potentially nasty surprises and eliminate or counter them in the system design, before they occur rather than after they occur.

2.6 Benefits of System Safety

When system safety is implemented at the start of design, the probability of mishaps is significantly reduced, along with a corresponding decrease in loss due to accidents/mishaps. Mishap prevention costs are generally less than mishap costs, therefore safety reduces overall total expected system costs. Conversely, if safety is not designed-in, mishap costs increase because mishap prevention is not implemented in the system design. When there is a late start in safety, there may be a delay in identification of safety design requirements, with the resulting possibility that design is well advanced before recognition of a hazardous condition.   At that point, it is much more expensive to implement a design fix, so there is a temptation to do nothing and take the risk. In an extreme case, while redesign is in progress other disciplines may be forced to cut back or delay the schedule, pending solution of the safety-related design problem.

The costs of a safety program are generally far less than accident and alteration costs in a laissez-faire program, which assumes safety as an incidental ingredient of design.  Skimping on safety funds, by providing a late start or intermittent effort, can save cents but waste dollars when an impending mishap occurs. The requirements for safe systems and reasonable system costs are compatible and are mutually supporting.

The results of system safety are often not visible when the SSP has been successful in preventing mishaps. This phenomenon tends to underrate and undervalue the benefits of the SSP. However, when mishaps do occur, system safety (or the lack thereof) becomes very visible. Some of the tangible and intangible benefits of system safety include:

- Preventing loss of life and/or serious injury
- Preventing system loss
- Preventing harm to the environment and the surrounding community
- Saving dollars and resources that might be lost from mishaps

---

[3] Reason, James, *Human Error*, Cambridge, Cambridge University Press, 1990.

- Maintaining a viable military force (by not losing systems, equipment and people to mishaps)
- Avoiding lawsuits and negative public opinion resulting from mishaps
- Meeting societal expectations for safety and risk

2.7 When Should System Safety Be Applied?

Every product and system should be made safe and proven to be safe before it is placed in the field for operational use. As such, system safety applies to every product and system for its entire lifecycle. Every organization and program should always perform the system safety process on every product, process or system. This is not only to make the system safe but also to prove and verify the system is safe. Safety cannot be achieved by chance; otherwise, failure will occur. One needs to remember: fail to plan = plan to fail. Safety cannot be achieved by chance. This concept makes obvious sense on large safety-critical systems, but what about small systems that seem naturally safe? Again, a system should be proven safe, not just assumed to be safe. A System Safety Program (SSP) can be tailored in size, cost and effort, depending upon the project. An SSP for a toaster system would be significantly smaller than an SSP for a motorcycle system, which would be smaller than for a skyscraper system.

System safety should especially be applied to complex systems with safety-critical implications, such as nuclear power, underwater oil drilling rigs, commercial aircraft, computer managed automobiles and medical systems. System safety should also be applied to integrated systems and systems of systems, such as automobiles within a traffic grid system within a highway system within a human habitat system.

The system safety process should particularly be invoked when a system has the potential to kill, injure or maim humans. It should always be applied as good business practice, because the cost of safety can easily be cheaper than the costs of not doing safety (i.e., mishap costs). When system safety is not performed, system mishaps often result, and these mishaps generate associated costs in terms of deaths, injuries, system damage, system loss, lawsuits and loss of reputation.

If the system safety process is not applied, or an inadequate process is performed, then some of the possible effects that may result include the following:

1) Safety becomes compliance-based rather than design-based.

2) Safety becomes process-based rather than hazard-based.
3) A return to the accident-based approach to safety results (i.e., fly-fix-fly).
4) Cheap workarounds are implemented.
5) Resulting problems may require product cancellation or complete redesign.
6) Fixes become costly when discovered and are implemented late in the program.
7) Mishaps and fixes may generate costly side effects.

When system safety is not applied, an absence of mishaps does not necessarily equate to a safe system, it only provides a false sense of safety.

2.8 The Cost of System Safety

The cost of safety is a two-edged sword – there is a cost for performing system safety and there is also a cost for not implementing system safety. An ineffective safety effort has a ripple effect; an unsafe system will continue to have mishaps until serious action is taken to correctly and completely fix the system. The cost of safety can be viewed as involving two competing components, the investment costs versus the penalty costs. These are the positive and the negative cost factors associated with system safety (i.e., safety as opposed to un-safety). When evaluating the cost of safety, both components must be considered because there is a direct interrelationship. In general, more safety effort results in fewer mishaps, and less safety effort results in more mishaps. This is an inverse correlation; as safety increases, mishaps decrease. There is a counterbalance here; it takes money to make safety increase, but if safety is not increased it will cost money to pay for mishap losses and to then make fixes that should have been done in the first place.

System safety should be viewed as an investment cost. Safety investment costs -- the actual amounts of money spent on a proactive safety program to design, test and build the system -- are an investment in the future. Because the system is designed to be inherently safe, potential future mishaps are eliminated or controlled such that they are not likely to occur during the life of the system. This investment should eliminate or cancel potential mishap penalty costs that could be incurred due to an unsafe system, thus saving money. One reason decision makers like to avoid the necessary investment cost of safety is because the results of the investment expenditure are usually not apparent or visible; they tend to be an intangible commodity (that is, until a mishap occurs). Penalty costs are the costs associated with the occurrence of a mishap or mishaps during the life of the system. Penalty costs should be viewed as the un-safety costs incurred due to mishaps that occur during operation of the system.

For example, consider the case where an SSP is conducted during the development of a new toaster system. Hazard analyses show that there are certain over-temperature hazards that could result in a toaster fire, which could in turn result in a house fire and possible death or injury of the occupants. The SSP recognizes that the risk of the new toaster system is too high and recommends that certain safety features be incorporated, such as an over-temperature sensor with automatic system shutdown. These safety features will prevent potential fires, along with preventing the penalty costs associated with the potential mishaps. However, the total number of house fires actually prevented by the new design might not ever be known or appreciated (thus safety has an intangible value).

Typically, the cost of safety is much less than the cost of not making the system/product safe, as the mishaps that may result can be quite expensive. Safety must be earned through the system safety process; it cannot be achieved by accident, chance or luck. Safety is not free, but it costs less than the direct, indirect and hidden costs of mishaps. The bottom line is that everything boils down to spending funds on performing system safety to produce a safe system versus not spending money (or not enough) and then paying for the various consequences of a mishap or mishaps. As with so many other things in life, either you're going to pay up front or you're going to pay on the back end for safety, and unfortunately, when you pay on the back end it's always more expensive.

2.9 Safety Liability

When mishaps occur they quite often stimulate lawsuits claiming the product or system was defective or unsafe in design. Product liability lawsuits are somewhat of a two-edged sword; on the one side there are unnecessary frivolous lawsuits, but on the other side of the sword there are valid lawsuits that are necessary in order to compel companies giving lip service to safety to start making safety a true priority.

Legal opinion has established that meeting safety standards is not irrefutable evidence that a product is safe. The fact that a particular product meets or exceeds the requirements of its industry is not conclusive proof that the product is reasonably safe. In fact, standards set by an entire industry can be found negligently low if they fail to meet the test of reasonableness.[4]

Codes and standards may serve as a floor, but not a ceiling, for safety liability. This means that something more than standards is required to develop a safe design and to provide evidence that a reasonable effort was made to develop a safe product/system. This is a major reason for performing a system safety process; system safety provides the tools and techniques for developing safe products/systems, along with supporting evidence. There is an economic advantage to hazard identification and elimination; when manufacturers design safety into their product there is less chance of experiencing mishaps and liability lawsuits. If it can be shown that a system safety program was not implemented, the defense against liability is significantly damaged.

## 2.10 Safety Culture

The safety culture of an organization is the resulting product of the individual and group values, attitudes, competencies and patterns of behavior that establish the organization's commitment to safety. Organizations with a positive safety culture are characterized by communications founded on trust, by shared perceptions of the importance of safety and by confidence in the effectiveness of preventive safety measures. It is important to understand that safety culture is a sub-set of the overall culture of the organization. It follows that the safety performance of an organization is greatly influenced and enhanced by its attitude toward safety and system safety.

A true safety culture is one in which every person in the organization recognizes their responsibilities to the organization and works to improve safety within the company and within the products produced by the company. There is a recognition that mishaps are not inevitable, and that mishaps can be prevented or acceptably reduced in risk.

A positive safety culture establishes an atmosphere where the employees are fully dedicated to safety and system safety and they receive management support in this endeavor. In a negative safety culture the opposite is true; safety commitment is strangled by the cynicism of some individuals, there is a fear of expending too much effort on safety, and there is a visible lack of management support. The safety of products and systems thrives when

---

[4] 63 Am. Jur. 2d Products Liability Section 381 (2011).

management actively establishes a positive safety attitude and environment. This is most effectively achieved when management makes safety a core value of the company. Leadership and commitment from the chief executive are required. Safety attitude and commitment flows from the top down, not the bottom up.

## 2.11 System Safety Standards

System safety is a process that is formally recognized internationally and used to develop safe systems in many countries throughout the world. MIL-STD-882 has long been the bedrock of system safety procedures and processes, and the discipline has grown and improved with each improvement in MIL-STD-882. Since the advent of MIL-STD-882, many other agencies, organizations and standards groups have developed their own variation on the system safety process.

Key system safety reference standards include the following:

1) MIL-STD-882, Standard Practice for System Safety, Original – 15 July 1969, Version D – 10 February 2000.
2) ANSI/GEIA-STD-0010-2009, Standard Best Practices for System Safety Program Development and Execution, 12 February 2009.
3) NAVSEA SWO20-AH-SAF-010, Weapon System Safety Guidelines Handbook, 1 February 2006.
4) System Safety Handbook: Practices and Guidelines for Conducting System Safety Engineering and Management, Federal Aviation Administration (FAA), 30 December 2000.
5) NPR 8715.3, NASA General Safety Program Requirements, 12 March 2008.
6) Air Force System Safety Handbook, July 2000.
7) MIL-HDBK-764, System Safety Engineering Design Guide for Army Materiel, 12 January 1990.
8) SAE/ARP-4754, Certification Considerations for Highly-Integrated or Complex Aircraft Systems, Aerospace Recommended Practice, 1996.
9) SAE/ARP-4761, Guidelines and Methods for Conducting the Safety Assessment Process on Civil Airborne Systems and Equipment, 1996.
10) IEC 61508, Functional Safety of Electrical/Electronic/Programmable Electronic Safety-Related Systems, Parts 1 – 7, September 2005.

# 3   SYSTEM SAFETY PROGRAM (SSP)

## 3.1 SSP Overview

The System Safety Program (SSP) is the heart of the system safety process; it drives and controls the entire process. The SSP is the joint combination of people and tasks that implement and execute the system safety process on a system or product development program. The three major aspects of the SSP are: 1) the core system safety process tasks, 2) ancillary safety tasks and 3) SSP interfaces. These primary components of the system safety process are depicted in Figure 3.1 for a typical system development project.

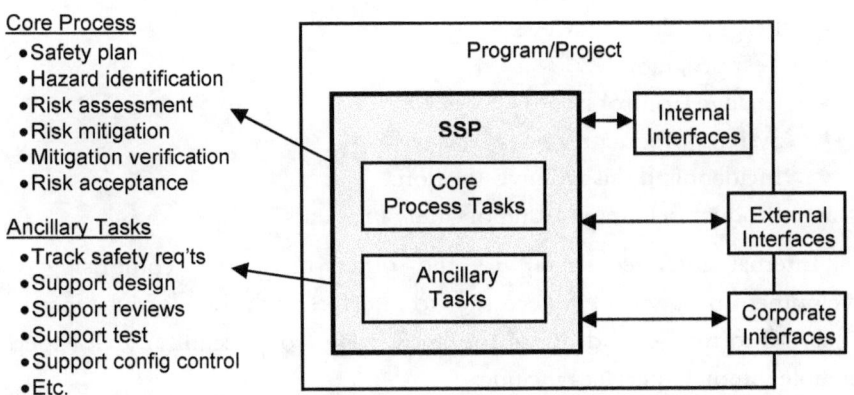

Figure 3.1 – Overall system safety Process

As shown in Figure 3.1, the system safety process primarily comprises the following basic elements:

- System safety program (SSP)
- Core process tasks
- Ancillary safety tasks
- Interfaces (internal, external and corporate)

The core system safety process consists of seven tasks that are always performed by a credible, effective and successful SSP (chapter 4 contains details). This core process is based on hazard identification, mishap risk and the risk management process for reducing risk. The *core* system safety process consists of the following basic elements:

1) Safety plan
2) Hazard identification
3) Risk assessment
4) Risk mitigation
5) Mitigation verification
6) Risk acceptance
7) Hazard tracking

Ancillary tasks refer to the many tasks that must be performed in addition to the core process. These tasks aid and abet the SSP, but by themselves alone will not guarantee a safe system. The ancillary tasks will vary depending on the type of system, the industry and industry standards involved and the safety criticality of the system. Example ancillary tasks include interfacing with, and supporting:

- Program design reviews
- Test program
- Change Control Board
- System/product readiness reviews
- Incident and mishap investigations
- Product field monitoring of safety incidents

Internal interfaces refer to the other disciplines comprising the development program. It is necessary to interface with these groups in order to share information and utilize the knowledge from technical area experts. Example internal interfaces include:

- Reliability engineering
- Systems engineering

- Quality
- Software engineering
- Design engineering (hardware, software, human factors)
- Testing
- Manufacturing

External interfaces refer to organizations or groups outside of the company that require interaction with the SSP, such as:

- Standards organizations
- Review boards
- Test laboratories
- Ancillary safety groups (e.g., test range safety)

Corporate interfaces refer to the interaction the SSP will have with corporate management. Corporate management must establish the company commitment, support and oversight for system safety. The system safety process is enhanced when the corporation makes safety a *true* core value for the company. This instills and supports a safety culture within the company.

## 3.2 SSP Description

An SSP is the combined set of people and tasks that implement and execute the system safety process on a development project or program. The SSP consists of an organization that performs the necessary system safety tasks and activities to realize the system safety objectives and obtain the necessary evidence of safety achievement. An SSP is the active, deliberate and intentional application of the system safety process by an individual, or group of individuals, skilled in that process. The system safety organization is a part of the larger organization developing a product or system. The SSP should be a distinct organizational entity appearing on the program organization chart, and it must be part of the program decision-making process. The SSP must have an organizational voice that can be heard by the program manager so that it is not muted by competing organizational goals. In general, the overarching objectives of the SSP are to:

- Manage and execute the system safety process
- Perform the core system safety elements
- Ensure that system design meets applicable safety requirements
- Ensure that all system hazards are identified and controlled
- Develop a system presenting minimal mishap risk

- Protect the system and its users, the public and the environment from mishaps
- Design-in safety into the overall system design or architecture

The intent of the SSP is to ensure that the system design meets applicable safety requirements and that all hazards associated with the system are identified and eliminated, or controlled in a manner consistent with program objectives, constraints and risk. The SSP also provides management visibility of safety risks inherent in the design and planned operations, and defines the process required for management to formally reduce and accept the system safety risks. Regardless of the type, size or complexity of a system, specific items are required to formulate an SSP. In order to exist as an entity, an SSP requires the following components, as a minimum:

- System safety organization
- Experienced system safety manager
- Experienced system safety staff
- Budget
- Program authority
- Program safety policy
- Contractual safety requirements
- System safety program  Plan (SSPP)
- Program management recognition and support
- A safety culture within the company

## 3.3 SSP Tailoring

The SSP can be tailored to fit the needs of the program, based on system size, complexity, safety criticality, funding, development schedule and other significant factors. Tailoring is recommended as there is no one-size-fits-all SSP; each SSP is unique and individual. The SSP is thoroughly planned and then documented in the System Safety Program Plan (SSPP) (see chapter 4 for detail on the SSPP).

## 3.4 SSP Scope

The scope of an SSP includes hardware, software, firmware and Human Systems Integration (HSI) for all system lifecycle phases. System safety requires a team effort whereby technical area experts should support the SSP and the system safety process. Since humans are a major component in most systems, human factors should be a major consideration in the system safety process. The system design must protect against potential safety-critical human errors, and it must not force the user or operator to unintentionally commit a safety error.

The scope of an SSP also includes the entire system lifecycle. A system lifecycle does not include legal standards such as statutes of repose limiting the liability to a certain number of years regarding systems designed, manufactured, or distributed. The system lifecycle involves the actual phases a system goes through from concept through disposal. The system lifecycle is analogous to the human lifecycle of conception, birth, childhood, adulthood, death and burial. The lifecycle of a system is generic and generally a universal standard. The system lifecycle stages are typically condensed and summarized into five major phases: Concept Definition, Development and Test, Production, Operation and Disposal.

It should be noted that in addition to the core system safety elements, an SSP also includes many program support tasks that are necessary, such as design reviews, technical review boards, system safety working groups, etc. These support tasks are ancillary to the core system safety process. The support tasks may vary, depending on the type, size and safety criticality of system.

## 3.5 SSP Organization

The system safety organization is responsible for performing the system safety tasks for the program; it is part of the larger program organization developing a product or system. The SSP organization should be a distinct organizational entity appearing on the program organizational chart with the responsibility and authority for the system safety function. The SSP organization should be part of the program decision-making process, with an organizational voice that can be heard by the program manager (PM) so that system safety is not unduly muted or constrained by competing organizational goals. The system safety organization is responsible for performing the tasks necessary to develop a system that presents minimal acceptable mishap risk, while considering competing factors and constraints, such as cost, schedule, technical complexity and mission expediency. It is the responsibility of the system safety manager to establish and manage the SSP organization. This requires a person knowledgeable and skilled in the system safety discipline and process.

## 3.6 SSP Staff

The SSP staff should comprise professional system safety engineers and analysts. Since the SSP staff will perform the hazard analyses and risk assessments, they must be experienced and knowledgeable in system safety, particularly the standards and guidelines that apply to the industry involved. Programs that do not use experienced system safety professionals usually do not perform thorough and correct hazard analyses and risk assessments. Experience has shown that system safety tasks require someone who can think in "failure space" as opposed to "success space" in order to properly identify and mitigate hazards. A highly effective company establishes qualifications for its SSP staff.

System safety is a specialized subject, just as are disciplines such as aerodynamics and structures. Every engineer cannot be expected to be an expert in all fields. An expert in hydraulics, for example, cannot be expected to be an expert in system safety and knowledgeable in all aspects of the system safety field. This is why a dedicated and qualified system safety engineer is required.

### 3.6.1 System Safety Management

System safety management is an element of program management which ensures the accomplishment of the system safety tasks, including identification of the system safety requirements: planning, organizing, and controlling those efforts which are directed toward achieving the safety goals; coordinating with other program elements; and analyzing, reviewing, and evaluating the program to ensure effective and timely realization of the system safety objectives.

### 3.6.2 System Safety Engineering

System safety engineering is an element of systems engineering that applies scientific and engineering principles to the timely identification of hazards and initiation of those actions necessary to prevent or control hazards within the system. It draws upon professional knowledge and specialized skills in the mathematical, physical, and related scientific disciplines, together with the principles and methods of engineering design and analysis to specify, predict, and evaluate the safety of the system.

# 4    CORE SYSTEM SAFETY PROCESS

## 4.1 Seven Core Steps

The system safety process conducts the intentional and planned application of management and engineering principles, criteria, and techniques for the purpose of developing a safe system. System safety applies to all aspects of the system and all phases of the system lifecycle. The core system safety process involves the following basic elements:

1) Safety plan
2) Hazard identification
3) Risk assessment
4) Risk mitigation
5) Mitigation verification
6) Risk acceptance
7) Hazard tracking

The seven required elements in this core system safety process are always performed by a credible, effective and successful system safety program (SSP). This core process is based on hazards, risk and risk mitigation. Identifying hazards and mitigating safety risk involves a dynamic closed-loop process. Figure 4.1 shows the core steps as they form a closed-loop process within the overall system safety program (SSP).

This core system safety process is a dynamic design safety methodology for two primary reasons: 1) it evolves or changes as the system design evolves, and 2) it is a closed-loop process where each step may be revisited as design data is updated until acceptable mishap risk is achieved. The core system safety process is a mishap risk management process, whereby safety is achieved through the identification of hazards, the assessment of hazard-mishap risk and the control or mitigation of hazards presenting unacceptable risk. Hazards are identified and continuously tracked until acceptable closure action is implemented and verified. This process should be performed in conjunction with actual system development, in order that the design can be influenced during the design process, rather than trying to implement costly design changes after the system is developed.

Figure 4.1 – Core SSP Process, Closed-Loop View

System safety, though involved in many different aspects of system/product development, is structured around the seven core elements that form the system safety process. These elements are building blocks that shape the foundation for the SSP. Figure 4.2 shows the core system safety elements and their interrelatedness. This viewpoint shows the core process as a sequence of tasks; however, in reality they are quasi-sequential steps, as the process has many iterations and interrelationships.

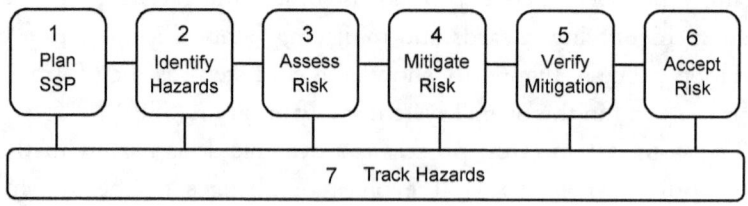

Figure 4.2 – Core SSP Elements, Task View

Note that the system safety process begins with Step 1, which plans and documents the entire process, which includes steps 2 through 7, plus it includes the plan for any required ancillary tasks. As Steps 2 through 6 are performed, their output is fed into the Hazard Tracking System (HTS), Step 6, for data retention and report generation. The six core steps are firmly established; however, some of the tasks performed within each step may vary slightly depending on the type of program and contracting involved. For example, on a military SSP the risk acceptance authority persons are mandated by DoD policy, whereas in a corporate SSP they may be delineated in corporate policy and look completely different.

The objective of the system safety process is to achieve acceptable mishap risk through a systematic approach of hazard risk management, involving hazard identification, hazard risk assessment, and hazard risk mitigation. Thus, system safety is a well thought out process that is planned, proactive, prudent and preventive in nature.

## 4.2 The Core Elements Defined

### 4.2.1 Element 1 – Safety Plan

Whereas the SSP consists of the actual people, activities and products involved in implementing the system safety process, the safety plan is the document that formally defines the SSP tasks, products, interfaces and milestones that will be required of the SSP organization. The plan also defines the scope of the safety program and decision criteria to be used by the program. The plan should cover the essential core components of an SSP, plus all of the relevant SSP support activities that will be required. The plan can be tailored to fit the needs of the program, based on system size, complexity, safety criticality, funding, development schedule and other significant factors. The safety plan must document the basic topics required in a management plan, including:

1) The necessary tasks for an effective safety program
2) The planned approach for task accomplishment
3) The tasks to be performed, task timing and task products
4) The qualifications of people to accomplish the tasks
5) The authority to implement tasks through all levels of management
6) The resources necessary to assure tasks are completed per the schedule.

The safety plan is typically recorded in a formal document known as the System Safety Program Plan (SSPP). This document defines and describes the

SSP and how it will be implemented; it is a plan of purpose, organization, action, methodology and schedule. It documents the management and engineering approach for applying the system safety methodology on a particular system or product development project. Years of experience have shown that safety can be more effectively designed-in when there is a roadmap for the entire process, and the SSPP is the roadmap. The principle of "fail to plan…plan to fail" applies here, with possible dire consequences.

A well-prepared and documented SSPP is the key to a successful SSP. The SSPP should be written to cover all aspects of the SSP. It should also be written to cover all phases where system safety work is to be performed, i.e., concept definition, design, test, deployment, operation, upgrade and disposal. The SSPP describes and formalizes the system safety management and engineering tasks and activities; it is the how-to document that provides the "what, when, why and who" for the SSP.

The depth, breadth and quality of the SSPP clearly reveal the value and importance placed on safety, and is an overall indication of management's commitment to system safety. It demonstrates how well system safety is understood by the safety manager, the program manager and the acquisition agency. The quality of the plan will indicate the quality of the SSP, which could also indicate relatively how safe the system will likely be when implemented. Although the SSPP format is not critical, it is essential that the SSPP contain the necessary and correct content. The basic program elements for an SSPP include the following, as a minimum:

- Introduction
- System Definition (short overview description)
- Scope boundaries of SSP
- System diagrams
  - System hierarchy table of all system elements
  - Physical and functional diagrams
  - Hardware, software and human interfaces
- Organization – SSP and overall program
- Roles and responsibilities
- Schedules – SSP and overall program
- Adopted safety precepts and principles
- Applicable safety requirements, guidelines and criteria
- Hazard identification approach (tied to the MEL)
- Hazard analysis methodologies to be utilized
- Depth of hazard analysis

- Risk assessment approach (including Risk Tables)
- Risk mitigation and verification approach
- Hazard tracking and closure approach
- Risk acceptance process and criteria
- Software safety approach
- Safety program compliance
- SSP products
- Safety data library
- SSP interfaces
- SSP support activities
- Resources
- Safety of Commercial Off-The-Shelf (COTS) items
- Safety Assessment Report (SAR)
- Program unique safety elements
- Review boards
- SSP audits
- System Safety Working Group (SSWG) support
- SSP Interfaces, such as with Configuration Management and requirements traceability

Although the content of each SSPP should contain the same basic elements, not all SSPPs are identical or necessarily look alike (nor should they). Each plan is unique to the particular type, size and safety criticality of system involved; it expresses the personality and management style of the organization developing the product. A tailored SSPP for a small project would normally look much different from a tailored plan for a large project. The SSPP for a non-safety-critical system would look different from one for an extremely safety-critical project.

The SSPP is a living document that is modified and updated as necessary during the life of the program. Updates may be required as each new development phase is entered to reflect new information and project changes. Also, during a particularly long development phase, the SSPP may require updating for various program-related reasons, such as design modifications, technology refresh, etc.

## 4.2.2 Element 2 – Hazard Identification

Hazard identification is the act of performing formal analyses to identify and evaluate hazards. Hazard identification typically tries to answer three questions:

1) What can go wrong that will lead to a mishap (hazard and its causal factors)?
2) What are the consequences (risk factor)?
3) What is the likelihood of occurrence (risk factor)?

Potential mishaps exist as hazards, and hazards exist within system designs. Hazards are actually inadvertently designed-in to the systems we design, build and operate. Sometimes this happens intentionally, but most often unintentionally. In order to perform hazard analysis, the analyst must first understand the nature of hazards. Hazards are predictable, and what can be predicted can also be eliminated or controlled.

Hazard identification is achieved through formal hazard analysis (HA) methods. The primary purpose of HA is to identify hazards and to obtain sufficient hazard data for risk assessments. HA is applied to hardware, software, functions, procedures and human tasks. HA can be applied at all stages of the system lifecycle. An HA becomes more detailed and accurate as more information about the system becomes available. Different HA techniques and approaches to hazard identification may be required at different stages of the system life cycle to ensure all types of hazards are identified. See chapter 9 for more details on basic hazard analysis methodology.

HA is the systematic examination of a system, item or product within its lifecycle to identify hazardous conditions, including those associated with human, product and environmental interfaces, and to assess their consequences to the functional and safety characteristics of the system or product. In order to design-in safety, hazards must be designed-out (eliminated) or mitigated (reduced in risk), which can only be accomplished through HA.

Since hazards are the primary focus of the core process, HA provides the basic foundation for system safety. HA is performed to identify hazards, hazard effects and hazard causal factors. HA is used to determine system risk, to determine the significance of hazards, and to establish design measures that will eliminate or mitigate the identified hazards. HA is used to systematically examine systems, subsystems, facilities, components, software, personnel, and their interrelationships, with consideration given to logistics, training, maintenance, testing, modification, and operational environments. In order to effectively perform hazard analyses, it is necessary to understand what comprises a hazard, how to recognize a hazard, and how to define a hazard. To develop the skills needed to identify hazards and hazard causal

34

factors it is necessary to understand the nature of hazards, their relationship to mishaps, and their effect upon system design. This skill is generally possessed only by experienced system safety practitioners.

MIL-STD882 recommends that the seven primary HA techniques of: Preliminary Hazard List (PHL), Preliminary Hazard Analysis (PHA), Subsystem Hazard Analysis (SSHA), System Hazard Analysis (SHA), Operating and Support Hazard Analysis (O&SHA), Health Hazard Assessment (HHA) and Safety Requirements/Criteria Analysis (SRCA) always be performed on each system development program. Due to system complexity, it typically requires more than one type of HA to identify all system hazards. Figure 4.3 contains a sample milestone chart showing the time period when each of these techniques is performed during the system lifecycle.

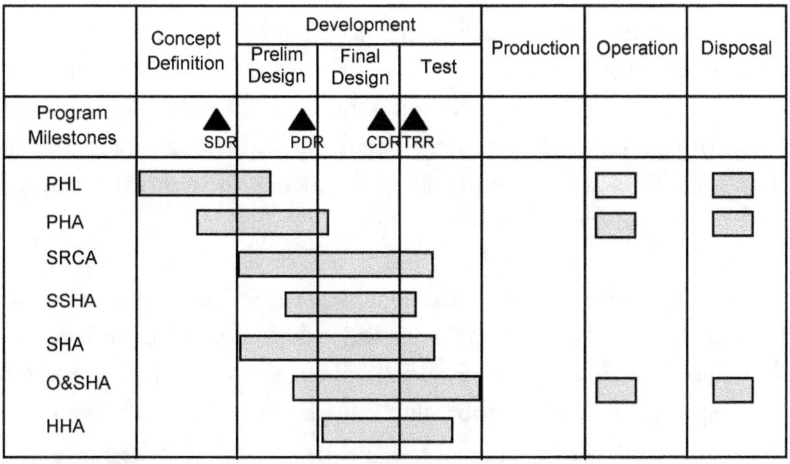

| | Concept Definition | Development | | | Production | Operation | Disposal |
|---|---|---|---|---|---|---|---|
| | | Prelim Design | Final Design | Test | | | |
| Program Milestones | ▲ SDR | ▲ PDR | | ▲ CDR ▲ TRR | | | |
| PHL | ███ | ██ | | | | ▭ | ▭ |
| PHA | | ███ | | | | ▭ | ▭ |
| SRCA | | ███ | ████ | | | | |
| SSHA | | | ███ | | | | |
| SHA | | ███ | ███ | | | | |
| O&SHA | | | ███ | ██ | | ▭ | ▭ |
| HHA | | | ███ | ██ | | | |

SDR-System Design Review; PDR-Preliminary Design Review; CDR-Critical Design Review; TRR-Test Readiness Review

Figure 4.3 – Program Milestone Diagram With HA Timing

Within the system safety discipline over 100 different HA techniques have been developed, some of which are unique, some of which are variants of others, some of which are useful and some of which are not useful. Each HA technique has a unique methodology, format and goals. It should be noted that there are many analysis techniques that are incorrectly used for HA, such as Failure Mode and Effects Analysis (FMEA). FMEA results can be used as input information for an HA, but the FMEA does not suffice for an HA because it does not thoroughly cover system hazard-mishap scenarios and it does not cover the combined effect of multiple simultaneous failures. Chapter 9 provides more detail on the hazard identification methodology, and Appendix A provides an example HA.

### 4.2.3 Element 3 – Risk Assessment

In the field of system safety, risk assessment involves identifying and assessing risks that can lead to mishaps. Potential mishap risk can be identified only by first identifying system hazards that create the risk. By assessing risks, priorities can be set for the allocation of resources to mitigate hazards and risk, and thereby minimize risk. Also, when the risk of a hazard is reduced through design safety measures, the residual risk is assessed to determine if it is acceptable or if additional safety measures are required.

Risk is a measure used to characterize the uncertainty associated with potential future events, in order that decisions regarding these events can be made today. Safety risk is characterized as "hazard risk" or "mishap risk", where risk is defined as: Risk = Likelihood X Severity. The likelihood factor is the likelihood of the hazard components occurring and transforming into the mishap. The severity factor is the overall consequence of the mishap, usually in terms of loss resulting from the mishap (i.e., the undesired outcome). Likelihood can be characterized in terms of probability, frequency or qualitative criteria, while severity can be characterized in terms of death, injury, dollar loss, etc. Both probability and severity can be defined and assessed in either qualitative terms or quantitative terms. Time is factored into the risk concept through the probability calculation of a fault event, for example, $P_{FAILURE} = 1.0 - e^{-\lambda T}$, where T = exposure time and $\lambda$ = failure rate.

*Hazard risk* is a safety metric characterizing the likelihood of a hazard occurring and transforming into a mishap, combined with the expected severity of the mishap described by the hazard. *Mishap risk* is the safety metric characterizing the likelihood of a mishap occurring, combined with the resulting severity of the mishap. It should be noted that hazard risk and mishap risk are really the same entity, just viewed from two different perspectives. Hazard analysis looks into the future and predicts a potential mishap from a hazard perspective. Mishap risk can be reached only by first determining the hazard and its risk. Since a hazard merely pre-defines a potential mishap, the risk has to be the same for both. Hazard risk exposes the potential mishap impact or threat presented by a hazard.

Risk assessment is the process of characterizing and ranking hazards based on risk. The purpose is twofold: 1) to determine which hazards present unacceptable risk, and 2) to prioritize hazards by risk level to determine which ones to eliminate or control first. This allows the program manager to know how and where to allocate resources to mitigate the risk in the most effective manner.

This task involves assessing the severity and probability of mishap risk for all identified hazards. It determines the potential negative impact of the mishap on personnel, facilities, equipment, operations, the general public, and the environment, as well as on the system itself. The steps performed in the risk assessment process are:

1) Identify the hazard
2) Identify and establish the hazard causal factors
3) Establish the likelihood of the causal factors occurring
4) Identify and establish the hazard's outcome severity
5) Combine the likelihood with the severity to determine the risk
6) Compare the computed risk with pre-established rating criteria
7) Take the necessary steps to eliminate/mitigate those hazards with unacceptable levels of risk

There are many different methods for rating the risk presented by a hazard and then determining if it is acceptable or not. Chapter 8 on Risk Theory describes this aspect in more detail, and Appendix B provides an example methodology.

## 4.2.4 Element 4 – Risk Mitigation

Risk mitigation is also known as hazard mitigation. It is the process of prioritizing risks, determining which risks must be mitigated first and then implementing design safety features to reduce the risk. An example of risk mitigation would be designing an automobile with air bags to reduce the risk of personal injury should an accident occur.

Risk mitigation involves establishing design safety methods needed to mitigate risk to an acceptable level. This involves translating safety design features into design requirements to support the system development process. Mitigation methods should be selected using the safety order of precedence. Mitigation methods should also be approved by all program stakeholders and technical areas experts.

System safety has an established preferred order of precedence for selecting design safety features, established by MIL- STD-882 (chapter 12 on

Design Safety Features contains more information). The Safety Order of Precedence (SOOP) for mitigating hazards by design safety is as follows:

1) Eliminate the hazard to the extent reasonably possible through alternate design means
2) If the hazard cannot be eliminated, reduce the hazard-mishap risk through:
   a) Design measures
   b) The use of safety devices
   c) The use of warning devices
   d) Special safety training and/or safety procedures

Risk mitigation methods typically target reducing the likelihood of a hazard occurrence and/or the reducing the severity of the hazard. For example, risk control measures such as double hulls and life-saving equipment are intended to reduce the consequences of a ship collision mishap. Equipping a ship with advanced redundant navigational systems and increased training to improve crew competency are methods to reduce the likelihood of a ship collision mishap. The keys to successful risk controls are that they be effective, specific to the hazards that create the risk and relatively easy to implement. This will increase the likelihood of successfully mitigating the hazard and the risk.

Since system safety is basically a safety-through-design process, this is a natural step to influence design via safety design requirements. Safety design requirements affect the design through the application of design guidance from prescribed sources and derived requirements. Prescribed sources include standards, specifications, regulations, design handbooks, and safety design checklists. Derived safety requirements are those that are generated specifically to ameliorate identified hazards not covered by prescribed sources. The hazard mitigation methods should be recorded in the appropriate hazard analysis and/or the hazard tracking system. Specific safety requirement numbers from the system specification documents should also be recorded.

4.2.5 Element 5 – Mitigation Verification

Mitigation verification involves ensuring that the risk mitigation methods have been implemented and are effective. This involves traceability of safety requirements, from design implementation through test verification. It also involves testing the design implementation to verify evidence of completeness and success.

This task involves assessing and verifying the effectiveness of the implemented safety design requirements through analysis, test or inspection. The residual mishap risk, as determined by test and/or audit/inspection, is recorded and documented. All test anomalies and software trouble reports should be reviewed by the safety engineer to determine if any new hazards identified during this process must be reported to the program manager. If testing to verify risk reduction is not required, then a safety audit, via physical inspection, is required to determine if all mitigation items are actually in the equipment/system. Note that regression testing is usually required for software that has been modified to confirm residual risk level. The results of verification should be recorded in the appropriate hazard analysis and/or the hazard tracking system. It should be noted that verification evidence is a key factor in the process – safety must be proven and verified through evidence.

### 4.2.6 Element 6 – Risk Acceptance

Risk acceptance should not be viewed as a way of not eliminating hazards and avoiding safety action. Risk acceptance is the process of ensuring that residual hazards are made adequately safe, that the risk is properly communicated and that the risk is accepted at the appropriate management level. It's a fact of system theory that not all hazards can be eliminated; some will persist as residual hazards. For example, unless gasoline can be eliminated from the automobile, certain hazards associated with gasoline will always exist. These residual hazards must be identified and reduced in risk to a point where mishap likelihood and/or severity is reduced to an extremely low (acceptable) level. Risk acceptance is not a simple clear-cut process. Hazards and risk exist regardless of our perception, knowledge or awareness of their presence. Hazards and risk do not care if we know about them or try to do anything about them.

Acceptable risk is the amount of potential mishap risk (i.e., danger) presented by an identified hazard that is allowed to persist without further risk reduction action. In system safety, risk acceptance is the formal process of accepting the risk presented by an identified hazard, thereby acknowledging its existence and the actions taken to control it. Risk acceptance involves communicating the risk to the proper program authorities and having them formally accept the risk for each identified hazard. If the risk acceptance authority believes the risk can be further reduced or an alternative method should be used, then the hazard will be sent back to the mitigation and risk assessment steps.

In order to know if a system is safe all the hazards and their attached risk must be known, and then the risk must be communicated and controlled until it is deemed acceptable. This requires an established risk management process, which includes a risk acceptance step. Some risk decisions are simple, such as deciding to cross a busy street, while others are more complicated, such as selecting a new car based on its unique safety features or choosing to implement a tri-redundant flight control system for aircraft safety. Sometimes, however, we are not even aware that we are participants in a risk acceptance situation, such as when a trucking company transports hazardous materials through our city streets, which could have a safety impact on our lives should a mishap occur.

Making the decision of risk acceptability is a difficult, yet necessary, responsibility of the system managing activity, system developer or system user. This decision is made with full knowledge that it is the user who is exposed to the risk, and the decision is often negotiated or tempered with competing factors such as cost, schedule and operational effectiveness. Risk acceptance is the culmination of a risk management process that results in a decision that the potential mishap risk presented by a hazard is known, understood and acceptable. Risk acceptance involves a decision-making process involving risk analysis and risk mitigation, in conjunction with program tradeoffs involving factors such as cost, schedule, design complexity, effectiveness, etc. The risk acceptance decision may involve a group effort; however, there is usually a specific decision authority responsible for the final decision, which requires a signature. When the decision authority decides to accept the risk, the decision must be coordinated with all affected organizations and then documented so that in future years everyone will know and understand the elements of the decision and why it was made.

One of the key components in determining if a hazard's risk is acceptable is a pre-established set of risk acceptance criteria. There are many models or methods for breaking down risk acceptance levels. Many different industries have their own individual methods already established. The U.S. Department of Defense has a risk acceptance process defined in MIL-STD-882, using the Hazard Risk Index (HRI) matrix concept, which is described in Appendix B.

### 4.2.7 Element 7 – Hazard Tracking

Hazard tracking is the process of systematically recording all identified hazards and the data associated with these hazards as they progress through the lifecycle of a hazard: identification, assessment, mitigation, verification, acceptance and closure. It is typically achieved through the use of a formal hazard tracking system (HTS). Hazard tracking is a basic required element of an effective system safety program (SSP). Hazard tracking encompasses the hazard analysis process, the mishap risk management process and the hazard mitigation process, and it ensures that no hazards are lost or overlooked.

A hazard tracking system (HTS) is a tool for formally tracking all identified hazards within a system. An HTS ensures that identified hazards are properly mitigated and closed, and that all related actions are recorded. An effective HTS actually establishes a process that facilitates hazard control through the steps of mitigation design, mitigation verification, risk acceptance and closure. Closed-loop hazard tracking is a basic required element of an effective system safety program (SSP). An HTS enforces a formal and systematic process that ensures identified hazards are resolved and also provides a historical record. It encompasses the hazard analysis process, the mishap risk management process and the hazard mitigation process. The HTS is sometimes referred to as a hazard log, hazard database or hazard tracking database (HTDB).

An HTS does not imply that a hazard is just passively stored in a database and then forgotten. Hazard tracking is a dynamic process in which the SSP takes positive steps to eliminate or mitigate the hazard and record all actions. Hazards are tracked from inception (identification) to closure, with focus on reporting and acceptance of the final residual hazard-mishap risk. Hazard tracking should be a "closed-loop" process, meaning that the review and mitigation process is repeated iteratively, until final closure of the hazard is achieved.

The primary objectives of an HTS include:

- Retain a record of all data and tasks associated with identifying and resolving hazards
- Provide a mechanism and discipline for tracking hazards from inception through closure
- Ensure that all identified hazards are adequately mitigated (i.e., none are lost)
- Meet the SSP requirement for hazard tracking and closure

The HTS provides a process for risk mitigation, as well as maintaining a complete record and history of every identified hazard. The hazard status is maintained as "open" until it has been verified that the appropriate safety requirements for eliminating or controlling the hazard have been implemented and proven successful through testing. Following successful verification and validation of the system safety requirements for a hazard, and acceptance of the hazard risk level, the hazard's status can be changed to "closed". Note that this requires documented evidence.

In an HTS, the database can be a manual or computerized system; however, it is highly recommended that an automated electronic database be utilized, particularly for medium and large system development programs. In addition, there are commercial electronic software HTS packages available that are already set up specifically for hazard tracking. An automated electronic database provides many advantages, such as:

- Hazard data entry is easy and efficient
- Data updates and changes are simple and efficient
- Capability exists to search for specific items based on different queries
- Capability exists to provide custom reports
- Capability exists to place on Network or Internet for access by many users
- Programming is not required if purchased
- Format can be utilized as a company standard for several different projects

Some basic considerations to address when designing, procuring and operating an HTS include the following:

- Rules for opening, monitoring and closing a hazard
- A hazard numbering scheme
- A standard format that remains consistent
- Rules for who can enter data into the HTS
- Rules for who can modify or remove data from the HTS
- The capability to generate various reports

4.3 Core Process Inputs and Outputs

Table 4.1 is an input-output table for each of the core process steps. This table describes the typical data required by each task and the output information generated by the task.

Table 4.1 – Core Process Input-Output Table

| Elements | Input Data | Output Data |
|---|---|---|
| Plan SSP | • SOW<br>• Program requirements<br>• Known safety requirements<br>• Applicable standards<br>• Program plans and schedules | • Safety plan<br>• HA method selection<br>• Risk criteria<br>• Software LOR criteria<br>• SSPP document |
| Identify Hazards | • Design data<br>• Operational data<br>• Historical data<br>• Problem reports<br>• Reliability data | • List of hazards<br>• Hazard records<br>• Hazard analyses |
| Assess Risk | • Hazard causal factors<br>• Failure rate data<br>• Risk rating criteria | • Risk assessment for each hazard<br>• Hazard risk record |
| Mitigate Risk | • Design options<br>• Safety policies and precepts<br>Reassessment of revised risk | • Safety design requirements |
| Verify Mitigation | • Test plans<br>• Test result reports<br>• Requirements verification | • Hazard closure if passed<br>• Hazard update if failed |
| Accept Risk | • Risk rating criteria<br>• Review by risk authority | • Risk acceptance letter |
| Track Hazards | • All identified hazards<br>• Data associated with hazards<br>• Risk assessments<br>• Safety requirements<br>• Verification results | • Database<br>• Reports<br>• Search results<br>• Safety case<br>• Lessons learned |

## 4.4 Evidence of Completion

In the core process there must be evidence of completion for each of the core elements in the system safety process. It is particularly imperative that conclusive evidence be provided that the mitigation methods were implemented and shown to be effective – safety must be shown through evidence. Table 4.2 provides a list of typical types of evidence provided for each of the core elements.

## Table 4.2 Evidence of Completion Table

| No. | Element | Evidence of Completion |
|---|---|---|
| 1 | Safety Plan | A formally documented System Safety Program Plan (SSPP). This should be a company document with an official company document number. The document should be approved by the appropriate management structure. |
| 2 | Hazard Identification | Formally documented hazard analyses. They should be documented in company documents with an official company document number and approved by the appropriate management structure. In addition, all identified hazards are entered into the Hazard Tracking System (HTS). |
| 3 | Risk Assessment | Formally documented risk assessments. They should be documented in a company document with an official company document number and approved by the appropriate management structure. The risk assessments can be included in the HA documents. In addition, all risk assessments can be attached to the applicable hazards in the HTS. |
| 4 | Risk Mitigation | Established safety design requirements that mitigate each applicable hazard. The safety requirements should be incorporated into the hazard analyses, the HTS and the system requirement specification documents. |
| 5 | Mitigation Verification | A formally documented verification plan documented in the SSPP. Mitigation verification results must be incorporated into the hazard analyses and/or the HTS. All design safety requirements must be verified for inclusion in the system and tested for effectiveness. |
| 6 | Risk Acceptance | A formal risk assessment process documented in the SSPP. It also includes the formal acceptance and closure of each individual hazard by the appropriate risk acceptance authority in a formal letter. |
| 7 | Hazard Tracking | A formal HTS, which must be maintained and kept available through the life of the system. |

# 5    SYSTEMS AND SYSTEMS LAWS

## 5.1 System Definition

System understanding is essential. We live within, and interface with, systems and systems-of-systems. Systems contain the factors that spawn hazards which can ultimately result in undesired events and mishaps. Understanding all of the factors involved with a particular system is necessary for the hazard identification and risk mitigation process. Because making the system safe is the goal of system safety, understanding system attributes and complexity is invaluable and necessary.

A system is an integrated composite of components that provides function and capability to satisfy a stated need or objective. A system is a holistic unit that is greater than the sum of its parts. Systems have structure, function, behavior, characteristics and interconnectivity. They vary in size, purpose, type, complexity and safety criticality. Modern day systems typically comprise people, products, processes and environments that together provide the capability to satisfy a stated need or objective.

Understanding system type and scope is very important in system safety and hazard analysis. The system type can indicate the safety criticality involved. The scope of the system boundaries establishes the size and depth of the system. The system limitations describe basically what the system can and cannot safely do. Certain limitations may require the system to include special design safety features. Every system operates within one or more different environments. The specific environment establishes what the potential hazardous impact will be on the system. System criticality establishes the overall safety rating for the system. A nuclear power plant system has a high consequence safety criticality rating, whereas a TV set, as a system, has a much lower safety criticality rating.

In short, a system is any group of interrelated, interacting and interdependent parts that form a complex and unified whole that has a specific function or purpose that is greater than that of the individual parts. If all the parts are not interrelated and interdependent, it is not a system, merely a collection of parts. Figure 5.1 characterizes the general aspects of a system as a unified whole.

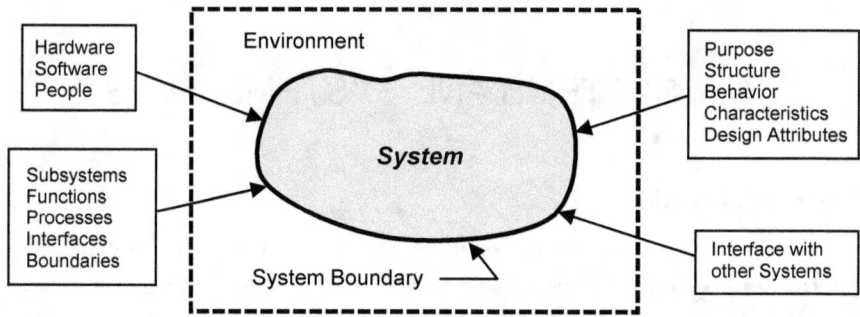

Figure 5.1 – System Characterization

Understanding the elements of a system is necessary in the performance of hazard analysis. A system typically comprises any combination of the following elements:

- Subsystems (sub-subsystems, units, assemblies, components)
- Hardware (electrical, hydraulic, structures, explosives, fuel, etc.)
- Software (program, segment, unit, module, logic, algorithms)
- People (operators, testers, maintainers, manufacturers)
- Processes (course of action, timing, material combining)
- Procedures (instructions, tasks, manuals, warning notes)
- Interfaces (hardware, software, documentation, communications)
- Functions (operations, modes, phases, tasks)
- Facilities (building, location, storage, transportation)
- Boundaries (physical, theoretical, limitations)
- Environment (weather, external equipment, temperature, vibration)

Two distinguishing characteristic of systems theory are that a) the whole is more than the sum of the parts, and b) what is best for the subsystems is not necessarily best for the overall system, and vice versa. The concept that the system is more than the totality of its components is referred to as synergy. As might be expected, these two system characteristics are important in safety analysis and in identification of hazards. The discipline of system safety must evaluate the subsystems as well as the system as a whole.

A system is a construct or collection of different elements that together produce results not obtainable by the elements alone. The elements, or parts, can include people, hardware, software, facilities, policies, and documents; all these things are required to produce systems-level results. A system has unique qualities, properties, characteristics, functions, behavior and performance. System safety works best at the system level because, as has been recognized, many hazards involve unique system interrelationships between the parts, rather than the parts in isolation. Figure 5.2 depicts the generic concept of a system. This diagram shows a system that comprises many subsystems, functions and interfaces between subsystems. System safety analysis involves evaluation of all system aspects, including subsystems, functions, interfaces, boundaries, and environments and the overall system itself.

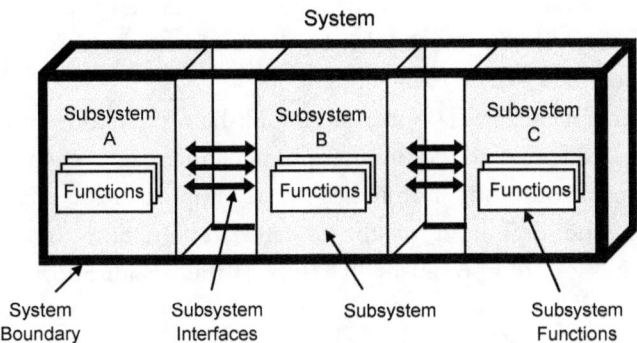

Figure 5.2 – System Model

5.2 Subsystem Definition

A subsystem is a subset of a system, a coherent and somewhat independent component of a larger system. A subsystem can include all of the same basic components that a system has, such as hardware, software, components, personnel, processes and procedures. A subsystem performs a specific function that contributes to accomplishing the system objective. Figure 5.3 displays the general aspects of a subsystem.

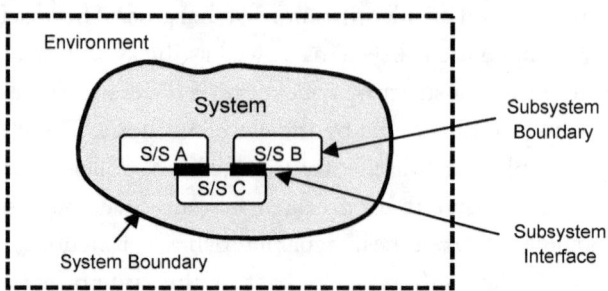

Figure 5.3 – Subsystem (S/S) Representation

## 5.3 System Hierarchy

Systems vary in size, shape, function, criticality and complexity. A system can be small, such as a toaster that consists of fewer than 50 parts. A system can be very large, comprising hundreds of subsystems, thousands of assemblies and millions of components, such as a commercial aircraft or a ship. Large complex systems can easily become overwhelming for human comprehension. In order to more easily understand a large system, the system is typically broken down or subdivided in a hierarchical manner into manageable pieces that can be easily understood.

The system hierarchy is typically established to define the system structure in an orderly and comprehensible manner. A system hierarchy establishes nomenclature and terminology that support clear, unambiguous communication and definition of the system, its functions, components, operations, and associated processes. It refers to the organizational structure defining dominant and subordinate relationships between subsystems, down to the lowest component/piece part level. Several closely related approaches have been established to formulate a system hierarchy. The Master Equipment List (MEL) is the systems engineering tool for exhibiting both system components and system hierarchy. The MEL is sometimes referred to as the Indentured Equipment List (IEL) because it shows hierarchy via indenture level of the components and functions as defined in MIL-STD-1629[5]. The MEL is also sometimes referred to as the Work Breakdown Structure (WBS) because it identifies both components and tasks in an

---

[5] MIL-STD-1629A, Procedures for Performing a Failure Modes Effects and Criticality Analysis, 24 Nov 1980.

indentured list structure as defined in MIL-HDBK-881[6]. Basically, the MEL is a list of all the systems, subsystems, units, assemblies and components in the major system, with each item in the list indented to reflect its hierarchy and ownership level. The indenture level also identifies or describes the relative complexity of assembly or function. The levels progress from the more complex (system) to the simpler (part) divisions. System design data and drawings will usually describe the system's internal and interface functions beginning at system level and progressing to the lowest indenture level of the system.

The following is a typical breakdown of successive indenture level groupings in the system hierarchy:

- System – an integrated set of subsystems that accomplish a defined objective
- Subsystem – an integrated set of assemblies, components, and parts which performs a cleanly and clearly separated function
- Assembly – an integrated set of components and/or subassemblies that comprise a defined part of a subsystem, e.g., the pilot's radar display console or the fuel injection assembly of an aircraft propulsion subsystem
- Subassembly – an integrated set of components and/or parts that comprise a well-defined portion of an assembly, e.g., a video display with its related integrated circuitry
- Component – a cleanly identified item comprising multiple parts, e.g., a cathode ray tube or the earpiece of the pilot's radio headset
- Part – the lowest level of separately identifiable items, e.g., a bolt

It is recommended that a system hierarchy table be established early in the program and used by the system safety practitioner in the performance of safety analyses. Figure 5.4 demonstrated the hierarchy of elements in a system along with a hierarchy table.

---

[6] MIL-HDBK-881, Work Breakdown Structure, 2 Jan 1998.

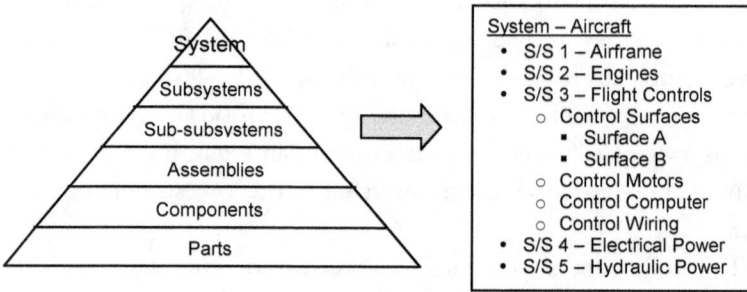

Figure 5.4 – System Hierarchy Representation

When performing hazard analysis, all of the system components must be considered to ensure a complete analysis. The following are some benefits gained from utilizing a system hierarchy table in safety analyses:

- Provides a complete list of equipment and functions
- Helps ensure that all of the system hardware and functions have been adequately covered by the hazard analyses
- Helps set the level of detail for a particular safety analysis
- Helps establish at what level in the hierarchy hazards should be identified
- Helps in determining at what system level risk should be assessed

5.4 System Types

The types of systems dealt with in system safety are typically physical, human-made objects comprised of hardware, software, user interfaces, and procedures. These types of systems include ships, weapons, electrical power, railroads, aircraft, etc., that are used for some specific purpose or objective. Table 5.1 provides some example systems, showing their intended purpose and some of the subsystems comprised within these systems. It is interesting to note that many of the systems comprise similar types of subsystems, which means that they may have similar types of hazards.

Table 5.1 – Example System Types

| System | Objective | Subsystems |
|---|---|---|
| Ship | Transport people /products Deliver Weapons | Engines, Hull, Radar, Communications, Navigation, Software, Fuel, Humans, HSI interfaces |
| Aircraft | Transport people /products Deliver Weapons | Engines, Airframe, Radar, Fuel, Communications, Navigation, Software, Humans, HSI interfaces |
| Missile | Deliver ordnance | Engines, Structure, Radar, Communications, Navigation, Software, HSI interfaces |
| Automobile | Transport people /products | Engine, Frame, Computers, Software, Fuel, Humans, HSI interfaces |
| Nuclear Power Plant | Deliver Electrical Power | Structure, Reactor, Computers, Software, Humans, Transmission Lines, Radioactive Material, HSI |
| Television | View Video Media | Structure, Receiver, Display, Electrical Power, HSI interfaces |
| Toaster | Browning of Bread | Structure, Timer, Electrical Elements, Electrical Power, HSI interfaces |
| Telephone | Provide communications | Structure, Receiver, Transmitter, Electrical Power, Analog Converter, HSI interfaces |

## 5.5 System Lifecycle

The system lifecycle -- the actual phases a system goes through from concept through disposal -- is analogous to the human lifecycle of conception, birth, childhood, adulthood, death and burial. The lifecycle of a system is very generic and generally a universal standard. The stages are typically depicted as the five major phases shown in Figure 5.5. All aspects of the system lifecycle can be characterized by one of these major categories or phases in this lifecycle model.

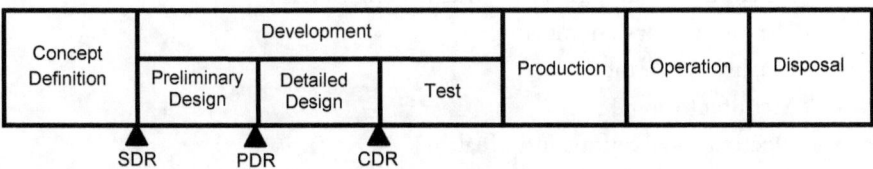

Figure 5.5 – System Lifecycle Phases

This is the standard traditional approach that has been in use for many years. The development phase is subdivided into Preliminary Design, Final Design and Test for more refinement. Under this model, each phase must be complete and successful before the next phase is entered, thus the system is developed in sequential stages. Three major design reviews are conducted for exit from one phase and entry into the next. These are the System Design Review (SDR), Preliminary Design Review (PDR) and Critical Design Review (CDR).

5.6 System Engineering Tools

In the process of developing large and highly complex systems, many tools are used by systems engineering and other related disciplines. Some systems engineering tools (described in chapter 10) that greatly aid the system safety analyst include:

- Simplified System Diagrams
- Functional Block Diagrams
- Master Equipment List (MEL)
- Reliability Block Diagrams (RBDs)
- Failure Mode and Effects Analysis (FMEA)

5.7 Systems Development Process

The task of developing large and highly complex systems requires the combined efforts of many different engineering groups and disciplines. Some systems engineering groups that support and interface with the system safety process include:

- Systems engineering
- Design engineering
- Software engineering
- Configuration management (design, requirements, change control)
- Test engineering
- Reliability, Maintainability and ILS engineering
- Human factors engineering
- Quality Assurance
- Manufacturing
- Occupational Safety and Health
- Airworthiness certification (when applicable)

5.8 Systems and Safety

As implied in the name, system safety involves *systems* and the many different characteristics and attributes associated with them. Systems have become a necessity for modern living, and each system spawns its own set of potential mishap risks. Systems can fail, malfunction and/or be erroneously operated. System safety engineering is the discipline and process of developing systems that present reasonable and acceptable mishap risk, for both users and nearby non-participants.

In order to achieve their desired objectives, systems are often forced to utilize hazardous sources in the system design, such as gasoline, nuclear material, high voltage or high pressure fluids. Hazard sources bring with them the potential for many different types of hazards, which if not properly controlled can result in mishaps. In one sense, system safety is a specialized trade-off between utility value and harm value, where utility value refers to the benefit gained from using a hazard source and harm value refers to the amount of harm or mishaps that can potentially occur from using the hazard source. For example, the explosives in a missile provide a utility value of destroying an intended enemy target; however, the same explosives also provide a harm value in the associated risk of inadvertent initiation of the explosives and the harm that would result. System safety is the process for balancing utility value and harm value through the use of design safety mechanisms. This process is often referred to as design-for-safety.

To design systems that work correctly and safely, a safety analyst needs to understand how things can go wrong and how to correct them. It is often not possible to completely eliminate potential hazards because a hazardous element can be a necessary component needed for the desired system functions, yet the hazardous element is what spawns hazards. Therefore, it is essential to identify and mitigate these hazards. System safety identifies the unique interrelationship of events leading to an undesired event in order that they can be effectively mitigated through design safety features. To achieve this objective, system safety has developed a specialized set of tools to recognize hazards, assess potential mishap risk, control hazards and reduce risk to an acceptable level.

A system is typically considered to be safe when it:

1) Operates without causing a system mishap under normal operation
2) Presents an acceptable level of mishap risk under abnormal operation, caused by faults and errors
3) Provides survival protection from foreseeable mishaps

Normal operation means that no failures or errors are encountered during operation (fault free), whereas abnormal operation means that failures, sneak paths, unintended functions and/or errors are encountered during operation. The failures and/or errors change the operating conditions from normal to abnormal. A system must be designed to operate without generating mishaps under normal operating conditions, i.e., under normal operating conditions (no faults) a system must be mishap-free. However, since many systems require the inclusion of various hazard sources, they are susceptible to potential mishaps under abnormal conditions, where abnormal conditions are caused by failures, errors, malfunctions, extreme environments and combinations of these factors. Hazards can be triggered by a malfunction involving a hazard source. During normal operation the design is such that hazard sources cannot be triggered. Normal operation relates directly to an inherent safe design without considering equipment or human failures. Abnormal operation relates to a fault-tolerant design that considers, and compensates for, the potential for malfunctions and errors combined with hazard sources.

## 5.9 System Laws

Several natural laws help describe the behavior of systems and help explain the various reasons why hazards exist within systems. There is no getting around these system laws; they will happen and they will shape the hazard risk presented by a system design. System safety must evaluate the potential impact of each of these system laws and determine if hazards will result, and if so, how the hazards can be eliminated or controlled to prevent mishaps. In other words, these system laws are hazard shaping factors that must be dealt with during product/process/system design in order to develop a safe system.

### 5.9.1 General System Laws

The following are important general system laws that help describe system uniqueness and complexity, which have an impact on both system safety and hazard analysis:

- Everything is a system.
- Every system is typically part of a larger system.
- The universe is systematized in a hierarchy of systems – smaller to larger systems.
- Many systems are integrated together (affecting size and complexity).
- Systems are typically complex, especially as they become larger.
- Systems are entities that comprise hardware, software, humans, rules, procedures, tasks and environments.

- Systems are designed to work in an intended manner for an intended purpose.
- System size and complexity directly impact total system understanding, reliability and safety.
- Safety and reliability are emergent properties of systems (that can be controlled).
- Systems react predictably to input; an expected input produces an expected output (normal operation).
- Systems react predictably to failures (thus, understanding failure modes and their effects is critical).
- Systems have no conscience

## 5.9.2 Reliability Laws

The following are general system laws directed specifically towards reliability, which help illuminate why systems fail and the reliability-safety relationship:

- A system is considered reliable when it works as intended for a specified time.
- A system is unreliable when it fails to work as specified or intended.
- Systems become unreliable due to failures, faults and errors.
- System size and complexity drive the number of possible components failures and failure modes.
- Reliability criticality is determined by failures that significantly impact reliability.
- System safety and reliability generally complement one another; however, they sometimes conflict.
- All things will eventually fail, due to aging, wearout, abuse, improper manufacturing, etc.
- Not all failures result in safety problems or hazards, whereas most failures impact reliability.
- When a component fails, the system design (and operation) effectively changes (safety must understand these changes).
- In large systems, failure and failure effects are complex.

## 5.9.3 Human Engineering Laws

Systems must be designed to account for, and counter, safety-critical human errors. The following are general human engineering laws directed specifically towards humans and human error, which affect safety:

- Humans will eventually fail and commit an error.
- Humans can be led into committing errors due to poor system design.

### 5.9.4 System Safety Laws

The following are general system laws directed specifically towards system safety, which help illuminate why hazards exist:

- Physical items will always eventually fail (unless they are repaired/replaced first).
- Humans do commit performance errors and always will.
- Systems must utilize components and functions that are naturally hazardous in order to achieve their basic purposes.
- Systems are sometimes designed with unintended functions that are not recognized.
- Systems are sometimes designed with hazardous sneak paths and integration flaws.
- Environmental factors can influence safe functioning of systems.
- Every system contains inherent potential safety vulnerability in some form.
- Safety vulnerability is characterized in terms of hazards, mishaps and risk.
- Hazards are a unique system attribute with identifiable and controllable components.
- Risk is the metric for measuring safety vulnerability.
- A system is safe when all mishap risk is known, understood, mitigated and judged as acceptable.
- System functions designed to operate in a specific manner have component failure modes that can result in inadvertent operation or incorrect operation of these functions.
- Hazards can exist without failures as the initiating mechanism (e.g., sneak circuit).

The conclusion drawn from these system laws is that we are surrounded by systems; systems are extremely complex and have a natural proclivity to fail. And, component failures are a major factor in many hazards, as is human error. Systems involve a conflict between man and machinery. These natural laws help to explain why hazards exist so abundantly within systems. It is the job of the system safety practitioner to use this information to identify hazards and then design to eliminate safety-related failure modes and human error modes, or design to counter them.

# 6 THE HAZARD-MISHAP RELATIONSHIP

## 6.1 Hazards and Mishaps

The term *hazard* is a common, well used term; most people feel they understand it and that its definition is intuitively obvious. In its colloquial usage the definition is fairly simple, straightforward and broad. However, in the field of system safety the technical definition of a hazard is a little more complex, narrower in scope and possibly less well understood. In the past, the terms *hazard*, *risk*, and *danger* have been used interchangeably and incorrectly in the public domain, resulting in an incorrect understanding of safety.

A hazard is typically defined as any potential condition that can cause injury, illness, or death to personnel; damage to or loss of a system, equipment or property; or damage to the environment. This is a good initial definition, but it leaves too much room for speculation and is difficult to use in exacting engineering applications. For example, using just this definition, it is commonly assumed that gasoline is a hazard. If gasoline is a hazard, then what is the risk involved? In attempting to mitigate hazard risk it becomes obvious that more information is required in order to effectively describe and understand a hazard. As will soon be learned, gasoline is a *hazard source*, but not a hazard.

For technical purposes, a hazard can be defined as the existence of a specific set of conditions that form the potential for a mishap event. In other words, a hazard is an existing system state that is dormant but has the potential to result in a mishap when the inactive hazard state components are actualized. A hazard is a potential mishap, while a mishap is an event that has occurred as a result of a hazard that has become armed and active. This more technical definition is necessary because in order to mitigate the risk presented by a hazard all of the components and parameters the hazard comprises must be identified and understood.

Understanding that a hazard is a *potential* mishap is essential to understanding safety, as there is a direct link between the two. The metric of risk can be derived from the hazard components. Risk can be changed (mitigated), but only when the hazard components are known and understood. Therefore, it is necessary to identify and understand the composition of all hazards within a system in order to understand and mitigate the risk before a mishap actually occurs.

Another way to visualize the direct link between a hazard and a mishap is to compare the definitions of each. A hazard is defined as any real or potential <u>condition</u> that <u>can cause</u> injury, illness, or death to personnel; damage to or loss of a system, equipment or property; or damage to the environment. A mishap is defined as an unplanned <u>event</u> or series of events <u>resulting in</u> death, injury, occupational illness, damage to or loss of equipment or property, or damage to the environment. These definitions show that the effect of each is the same – death, injury, etc. A mishap is an event resulting in loss and a hazard is a condition that can cause loss, thus, a hazard is a condition that can cause a mishap event.

A hazard is a physical entity that characterizes a potential mishap; it is a condition that is prerequisite to a mishap; it is a blueprint for a mishap. In order for a hazard to exist, three required components are necessary, which form a Hazard Triangle (chapter 7). A mishap is the result of an armed and actuated hazard. Hazards are the result of hazardous system components, poor design and/or inadequate design foresight.

A hazard and a mishap are *before* and *after* states linked by a transition mode. The transition mode arms and activates the hazard. This concept leads to the principle that a hazard is the precursor to a mishap; a hazard defines a potential event (i.e., mishap), while a mishap is the event occurrence. This relationship between a hazard and a mishap are depicted in Figure 6.1.

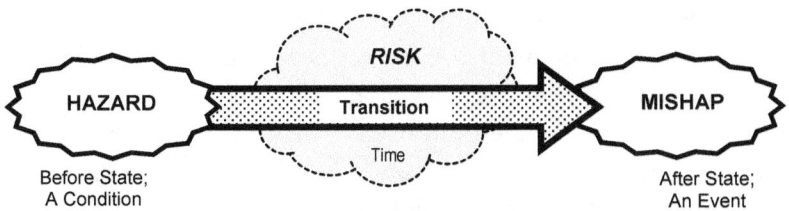

Figure 6.1 – Hazard / Mishap Relationship

A hazard and a mishap are two separate states of the same phenomenon, linked by a state transition. A hazard is a "potential event" at one end of the spectrum (before state), that transforms into an actual event at the other end of the spectrum (after state). The state transition results from the occurrence of the hazard Initiating Mechanisms. A hazard is a latent condition that exists in a system, that when activated results in a mishap. Mishaps are the immediate result of actualized hazards. The state transition from a hazard to a mishap is based on two factors: 1) the unique set of hazard components involved and 2) the mishap risk presented by the hazard components. The hazard components are the items comprised by a hazard, and the mishap risk is the probability of the mishap occurring and the severity of the resulting mishap loss.

Hazard/Mishap risk is a fairly straightforward concept, where risk is defined as:

$$\text{Risk} = \text{Likelihood } \textbf{x} \text{ Severity}$$

The mishap likelihood factor is the likelihood of the hazard components occurring and transforming into the mishap. Likelihood can be expressed in terms of probability, failure rate or qualitative ranges. The mishap severity factor is the overall consequence of the mishap, usually in terms of loss resulting from the mishap (i.e., the undesired outcome). Severity can be defined and assessed in either qualitative terms or quantitative terms. Time is factored into the risk concept through the probability calculation of a fault event, for example, $P_{FAILURE} = 1.0 - e^{\lambda_T}$, where $T$ = exposure time and $\lambda$= failure rate.

Risk is an important component of the hazard-mishap relationship. Risk is a measurement that rates the overall safety significance (or danger) of a hazard. Risk is a measure that tells safety analysts how likely, or how often, the hazard will transform into a mishap. The risk measure also tells the safety analyst how severe the final mishap outcome is likely to become.

Risk management is an important tool for making critical decisions, and, in some cases, meeting regulatory requirements. In system safety, risk management is a process for the identification of hazards and their causes, determining the consequences of the hazards, calculating the probability of their occurrence and determining whether the risk is acceptable or if corrective actions are needed to make the risk acceptable. Hazard risk management is a key element of the system safety process.

However, risk management must not negate the system safety objective of eliminating hazards. No hazard should be accepted when that hazard can be reasonably eliminated or reduced in risk via design safety measures. Risk should be accepted only when the benefits outweigh the potential mishap damages, losses and costs. Risk information, risk knowledge and risk management are not an excuse for not eliminating hazards when feasible.

# 7   HAZARD THEORY

7.1 Hazard Description

A hazard is typically defined as a potential condition that can cause a mishap resulting in undesired consequences. From a more technical perspective, a hazard can be described as the existence of a specific set of system conditions that together create the possibility of a potential mishap; a hazard is an existing dormant system state, which has the potential to result in a mishap when the dormant hazard components are actualized. A hazard is a potential, while a mishap is an actual. A hazard can be thought of as a scenario comprising specific dormant causal factors, that when activated, result in a mishap.

7.2 Why Hazards Exist

Hazards occur in a system for one simple reason – the system must utilize, or interface with, a hazard source in order to achieve its intended goals. A mishap occurs for two reasons – a hazard exists and the hazard was not eliminated or properly mitigated.

Hazards are created because of the need for hazardous sources in the system, or they must interface with hazard sources, coupled with the fact that eventually everything fails, and these failures can unleash the undesired effects of the hazard source. Hazards also exist due to the need for safety-critical system functions, coupled with the potential for failures and human error within these safety-critical functions. Hazard creation can be summarized by the following factors, which can occur singularly or in combinations:

- The use of hazardous system elements (e.g., fuel, explosives, electricity, velocity, stored energy).
- The system interfaces with hazard sources (e.g., fuel, electricity).
- Operation in hazardous environments (e.g., flood zones, ice, heat).
- The need for hazardous functions (e.g., aircraft fueling, welding).
- The use of safety-critical functions (e.g., flight control, arming).
- The inclusion of (unknown) design flaws, errors and sneak paths.
- The potential for hardware wear, aging and failure.
- Inadequacy in designing to tolerate critical failures.

## 7.3 Hazard Enigma

A hazard is as *a potential condition that can cause a mishap, resulting in undesired consequences.* There is a direct link between a hazard and a mishap; a hazard being the precursor to a mishap. The definition of a mishap is fairly well understood, it is an event that has occurred with undesired outcomes, such as death, injury and/or damage. An industry problem is that the definition and concept of a hazard is not well understood or precisely defined, especially for technical purposes. For example, what specifically is a "potential condition" and how can it be understood and represented in order to consistently identify hazards? A condition can be many different things to different people. In addition, a hazard appears to be an *actual* existing condition with *potential* consequences. In order to fully understand, appreciate, recognize and measure hazards a more rigorous model based definition is needed. It is no longer acceptable to leave the hazard definition open to individual interpretation.

A hazard is a real and existing condition with a potential undesired mishap outcome. It is recognized within the safety community that a hazard essentially has three states: dormant, armed and active. It is initially in an existing dormant condition, which is armed when it is activated by certain causal factors, and after becoming armed it transitions to the active state, which is the mishap event. The transition from armed to active may be immediate or it may take some length of time.

Hazard identification can no longer be an intuitive process, it must be a scientific process based on critical thinking. Several hazard models have been proposed; unfortunately many of them do not provide a consistent description of what actually comprises a hazard when it is in the *condition* state.

It is necessary to understand the constituent components of a hazard in order to consistently recognize them and evaluate their risk and mitigation. A *condition* is defined as a state at a particular time; a *state* is defined as the way something is with respect to its main attributes (i.e., a situation). This infers that a condition is a concrete situation (or scenario) that can be clearly specified, with precise parameters.

Understanding that a hazard is a *potential* mishap, with specific causal factors, is essential to understanding safety. The metric of risk can only be derived from the hazard causal factors. Risk can be changed (mitigated), but only when the hazard components are known and understood. Therefore, it is necessary to identify and understand the composition of all hazards, within a total system environment, in order to understand and mitigate the risk before a mishap actually occurs.

The HS-IM-TTO hazard model makes the most logical sense in defining what comprises a hazard. This model breaks the hazard *condition* into three recognizable aspects or components: a hazard source (HS), an initiating mechanism (IM) and a target-threat outcome (TTO). *All hazards require these three components in order to exist.* This model supports the dormant hazard state, the causal factors that cause the hazard to become armed and the final outcome to expect when the hazard becomes active and transitions to a mishap. The HS is the element that causes the basic danger; the IM is the element that arms and activates the hazard; the TTO is the expected outcome severity when the mishap occurs. The hazard *condition* is a scenario that could result in a mishap, which is characterized by the HS-IM-TTO constituent components. A hazard description must contain the complete scenario context with all three of the required hazard components identified.

There is a lot of confusion about hazards, for example, one hazard model states that a hazard can be caused by other contributory hazards. This suggests a multi-hazard link where there is an initiating hazard and then contributory hazards. This is not a useful (or correct) concept. A hazard can, however, be caused by one or more causal factors (trigger events). Also, different hazards can be very similar with slightly different outcomes due to a slight change in one of the causal factors. This situation could be called a family of hazards. Trigger events are the hazard initiating mechanisms, such as failures, human error, etc. In the HS-IM-TTO hazard model, the IM component accounts for all causal factors, whether or not they are referred to as initiating and contributory events (but they are not referred to as additional hazards). A hazard contains *only* the minimal causal factors that will cause it to occur. If additional *extra* causal factors can cause the hazard, then they form a separate, but similar, hazard. Bottom line – one hazard … one mishap … one minimal set of causal factors.

## 7.4 The HS-IM-TTO Hazard Model

The HS-IM-TTO hazard model is the most logical for defining what comprises a hazard. In order for a hazard to exist, three hazard components must be present to form the hazard: 1) the Hazard Source (HS) which provides the basic source of danger, 2) the potential Initiating Mechanisms (IM) that will transition the hazard from an inactive state to a mishap event, and 3) the Target-Threat Outcome (TTO), or consequences, that will result from the mishap event.[7] The existence of a hazard requires the existence of these three components as a prerequisite. A hazard is an existing potential condition (inactive) that will result in a mishap when actualized. The hazard condition is a potential state formed by the Hazardous Source (e.g., energy) and the potential Initiating Mechanisms (e.g., failures) that will transform the hazard into a mishap and the hazard Outcome. The hazard Outcome is predicted in the expected mishap Target (e.g., personnel) and the expected mishap Threat (e.g., death or injury). These components are necessary to assess the risk and to know where and how to mitigate the hazard. Table 7.1 describes these components in more detail.

Table 7.1 – Hazard Components

| Hazard Component | Function or Purpose | Examples |
|---|---|---|
| Hazard Source (HS) | This is the element that provides the basic source of danger. Without it there would likely be no hazard. | • Energy sources<br>• Safety-critical functions<br>• Adverse environments |
| Initiating Mechanism(s) (IM) | These are the initiators or mechanisms that cause the mishap event to occur. | • Hardware failures<br>• Human errors<br>• Bent connector pins |
| Target-Threat Outcome (TTO) | This describes the potential outcome when the hazard becomes an actual mishap. There has to be a potential target and a threat to that target in the form of consequence and severity. | • Death or injury threat to humans<br>• Damage threat to system or product<br>• Damage threat to the environment |

---

[7] Adapted from work by Pat Clemens as described in System Safety Scrapbook, sheet 98-1, *Describing Hazards*, 1998, Sverdrup Corp. and work by J. Letos in Boeing document D180-28993-302, *Introduction to System Safety Analysis Process*, 1995, page 22.

A more technical hazard definition is: A hazard is a set of dormant conditions, which consist of a Hazard Source, an Initiating Mechanism and a Target-Threat Outcome, which leads to a mishap when the Initiating Mechanism is actualized. A hazard is a physical entity that characterizes a potential mishap. A hazard is a condition that is prerequisite to a mishap, i.e., it is a blueprint for a mishap. When a hazard exists it forms a Hazard Triangle.

Basically, a hazard exists as a result of a system component being present in the system that presents safety vulnerability (i.e., the HS), combined with a system design that poorly tolerates various failure mechanisms that can affect the HS component. The actual amount of risk presented by the hazard is a function of how the design responds to the HS component when it is subjected to factors such as: hardware failures, human errors, sneak electrical paths, software errors, incorrect interfaces, etc. In order to mitigate a hazard, the hazard must be recognized and understood, and then the influence factors appropriately modified. Table 7.2 provides some example items and conditions for each of the three hazard components.

Table 7.2 –Hazard Component Examples

| Hazard Source (HS) | Initiating Mechanism (IM) | Target/Threat Outcome (TTO) |
|---|---|---|
| • Ordnance <br> • High pressure tank <br> • Fuel <br> • High voltage | • Inadvertent signal; RF energy <br> • Tank rupture <br> • Fuel leak and ignition source <br> • Touching an exposed contact | • Personnel; Explosion; Death/Injury <br> • Personnel; Explosion; Death/Injury <br> • Personnel; Fire; Death/Injury <br> • Personnel; Electrocution; Death/Injury |

To demonstrate the hazard HS-IM-TTO model, consider a detailed breakdown of the following example hazard: "Worker is electrocuted by touching exposed contacts in electrical panel containing high voltage." Figure 7.1 shows how this hazard is divided into the three necessary hazard components to validate the hazard. Note in this example that all three hazard components are present and can be clearly identified. In this particular example there are actually two IMs involved. The TTO defines the mishap outcome, while the combined HS and TTO define the mishap severity. The HS and IM are the hazard causal factors that are used to determine the mishap probability. If the high voltage component can be removed from the system, the hazard is eliminated. If the voltage can be reduced to a lower less harmful level, then the mishap severity is reduced and the hazard is mitigated to a lesser level of risk.

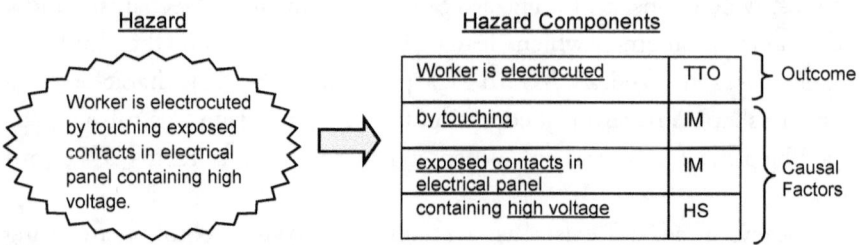

Figure 7.1 – Example of Hazard Components

A hazard is an existing potential condition (inactive) that will result in a mishap when actualized. The hazard condition is a potential state formed by the Hazardous Source (e.g., energy) and the potential Initiating Mechanisms (e.g., failures) that will transform the hazard into a mishap and the hazard Outcome. The hazard Outcome is predicted in the expected mishap Target (e.g., personnel) and the expected mishap Threat (e.g., death or injury).

Figure 7.2 depicts the S-M-TTO hazard model with more detailed examples of possible system factors that could produce each of the S-M-TTO categories. This example list of factors is not complete or exhaustive, but it becomes apparent that the list of HS factors can be very extensive. There are many HSs in the world, and in systems designs, that must be dealt with via the system safety process in order to produce safe product and systems.

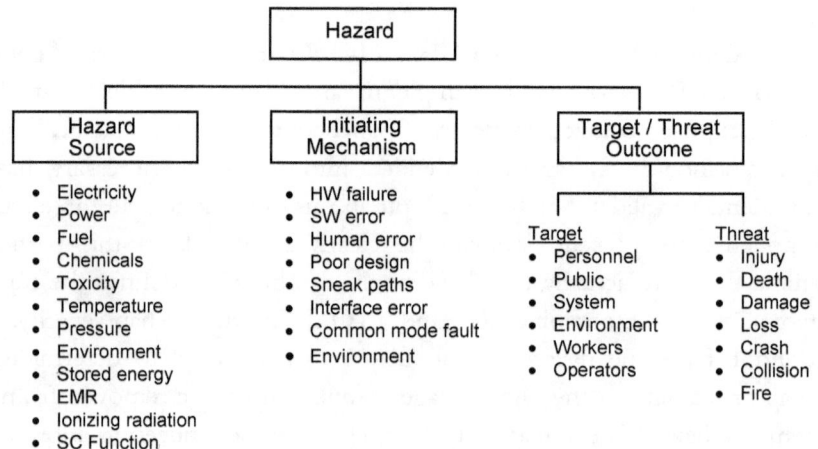

Figure 7.2 – Expanded Hazard Model

7.5 Hazard Triangle

The three required components of a hazard form what is known as the Hazard Triangle, which is illustrated in Figure 7.3. The Hazard Triangle conveys the idea that a hazard consists of three necessary and coupled components, each of which forms the side of a triangle. All three sides of the triangle are essential and required in order for a hazard to exist (i.e., HS, IM and TTO). Remove any one of the triangle sides and the hazard is eliminated because it is no longer able to produce a mishap (i.e., the triangle is incomplete). For example, remove the human operator from an aircraft, and the hazard "aircraft crashes resulting in pilot death" is eliminated because the TTO is removed. Reduce the probability of the IM side and the mishap probability is reduced. Reduce an element in the HS or the TTO side of the triangle and the mishap severity is reduced.

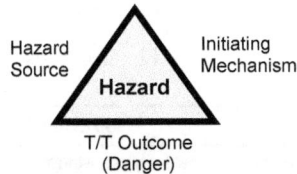

Figure 7.3 – Hazard Triangle

A hazard can be eliminated by eliminating any one of the three basic components of a hazard (HS, IM or TTO) because this breaks the hazard component coupling. Practically speaking, however, hazards are primarily eradicated by eliminating the HS component. Hazards are predominantly mitigated in risk by reducing the probability of the IM. Hazard risk can also be mitigated by reducing the effective danger of the HS, by protecting the Target or by reducing the amount of the Threat. It should be noted, however, that most hazards are mitigated by reducing the actuating probabilities of the IM. Changes in the probability and/or severity of a hazard modify the risk of the hazard. The Hazard Triangle concept is useful in determining if a conceptualized hazard meets the necessary criteria and in determining where to mitigate a hazard. It also demonstrates that when a hazard is mitigated it is not eliminated (a common error of safety beginners), because all three sides are still present. It's interesting to note that these mitigation methods correlate appropriately with the Safety Order of Precedence.

## 7.6 Good Hazard–Bad Hazard

Correctly describing the hazard is a very important aspect of hazard theory and hazard analysis. The hazard description must contain all three components of the hazard (Hazard Source, Initiating Mechanism, and Target/Threat Outcome). The hazard description should also be clear, concise, descriptive and to the point.

If the hazard description is not properly worded and clearly understood, it will be difficult for anyone other than the original analyst to completely understand the hazard. If the hazard is not clearly understood, the concomitant risk of the hazard may not be correctly determined. This can lead to other undesirable consequences, such as spending too much time mitigating a low-risk hazard, when it was incorrectly thought to be a high-risk hazard. Table 7.3 shows some good and poor examples of hazard descriptions.

Table 7.3 – Example Hazard Descriptions

| Poor Examples | Good Examples |
|---|---|
| Broken glass | Worker accidentally breaks glass window and severely cuts himself from the broken glass. |
| High voltage | Worker is electrocuted by touching exposed contacts in electrical panel containing high voltage. |
| Gasoline | Automobile is hit from the rear by another auto, causing the fuel system to rupture; spilled fuel is ignited, resulting in fire that severely injures occupants. |
| Repair technician slips on oil. | Overhead valve V21 leaks oil on walkway below, spill is not cleaned, repair technician walking in area slips on oil and falls on concrete floor, causing serious injury. |
| Signal MG71 occurs. | Missile Launch signal MG71 is inadvertently generated during standby alert, causing inadvertent launch of missile and death/injury to personnel in the area of impacting missile. |
| Round premature | Artillery round fired from gun explodes or detonates prior to safe separation distance, resulting in death or injury to personnel within safe distance area. |
| Ship causes oil spill | Ship operator allows ship to run aground, causing catastrophic hull damage, causing massive oil leakage, resulting in major environmental damage. |

Note that the good description examples contain all three elements of a hazard (HS, IMs and TTO). It should also be noted that the poor hazard description examples are primarily hazard sources only. It should be emphasized that occasionally the HS and IM components may overlap or their distinction is blurred. In these cases it is important that all the components are identified, and divisive arguing should not be spent on whether or not they are labeled specifically as HS or IM factors.

## 7.7 Hazard Context

Hazards must be described in a proper and complete systems context. This context must explain the complete system scenario and causal factors, which include the Hazard Source, Mechanism and Target/Threat Outcome. Do not abbreviate the hazard description or use program-unique lingo, assuming the reader fully understands the text. Describe the hazard scenario in a meaningful context that fulfills the three sides of the Hazard Triangle. For example, "fuel" is not a hazard, it is a hazard source. "Fuel leak occurs due to *xxx* and an ignition source occurs due to *xxx*, leading to fire and system loss" is a hazard. In this example, the actual causes of the fuel leak and the ignition source must be included. If the system context of the hazard is not fully described, then the hazard causal factors and hazard risk cannot be established, because a risk calculation requires something concrete from which to derive likelihood and severity.

# 8   RISK THEORY

8.1 Risk Concept

The concept and reality of risk has been around for some time. There are many different types of risk, such as: safety risk, hazard risk, mishap risk, schedule risk, cost risk, investment risk, product risk, sports risk, etc. Risk also involves many different viewpoints, such as: perceived risk, real risk, individual risk, group risk, societal risk, high risk takers, low risk takers, risk aversion, etc. On the surface, risk appears to be a very simple concept; however, risk can easily become very complex due to all the types, factors, possibilities and considerations involved. Risk and risk management are not just safety concepts; they are used in many different fields, such as finance, project management and health care, just to name a few. Risk is not about the present, it is about the future. Risk deals with uncertainty and future outcomes.

Risk is a vector value combining event likelihood with event outcome. It is a metric expressing the expected value of a future event based on the parameters creating the potential event. It expresses the likelihood of a potential gain/loss from a given decision. Risk involves three parameters: a) a potential future event, b) the likelihood of the event occurring and c) the potential consequences from the event when it occurs. Each of these aspects involves an element of uncertainty. Risk is defined as the product of the event likelihood and the event outcome, where the outcome can be either a positive or negative consequence, depending on the event. Risk outcome is the final expected result of the future event, given that it occurs. Risk outcome can be a threat or an opportunity; however, in system safety it is treated as a threat of loss, damage, death, injury or any combination of these outcomes. Safety risk is a method for quantifying danger and the uncertainty involved.

Risk is a measure of the uncertainty associated with future potential events, in order that decisions regarding these events can be made today. Risk decisions can result in either negative or positive outcomes. Risk management is a tool used to make decisions in the present that will help to produce a desired outcome in the future. Risk is an intangible quality; it does not have physical or material substance (a mishap does, but not risk). It is a future-value concept with some quantifiable metrics, likelihood and severity, that characterize the future event. In system safety, risk is a measure of the future event, where the event is an expected mishap. Risk likelihood can be characterized in terms of probability, frequency or qualitative criteria, while risk severity can be characterized in terms of death, injury, damage, dollar loss, etc. Recognizing that a hazard is the precursor (or blueprint) of a mishap, safety risk is the common denominator between the hazard and a mishap, and also the measure of the relative threat presented by a hazard. Risk is the metric characterizing the amount of danger presented by a hazard.

Safety risk is sometimes stated two different ways by individuals with different backgrounds, which can lead to confusion. The two types of safety risk terms in use are:

- Hazard risk – the safety metric characterizing the amount of danger presented by a hazard, where the likelihood of a hazard occurring and transforming into a mishap is combined with the expected severity of the mishap outcome predicted by the hazard.
- Mishap risk – the safety metric characterizing the amount of danger presented by a potential mishap, where the likelihood of the mishap's occurrence is combined with the resulting severity of the mishap. Mishap risk likelihood defines the likelihood of the mishap occurring, while mishap risk severity defines the expected final consequences and loss outcome expected from the mishap event. The mishap likelihood and severity can only be computed from the information contained in the hazard description, when the mishap is being predicted in advance.

It should be noted that hazard risk and mishap risk are really the same entity, just viewed from two different perspectives. Hazard analysis looks into the future and predicts a potential mishap from a hazard perspective. A potential mishap can only be recognized and understood in terms of a hazard; therefore, mishap risk can only be reached by first determining the hazard and then evaluating its risk in terms of a mishap from the hazard causal factors. Since a hazard merely pre-defines a potential mishap, the risk has to be the same for both of them. Figure 8.1 depicts the hazard–mishap risk relationship and their common link.

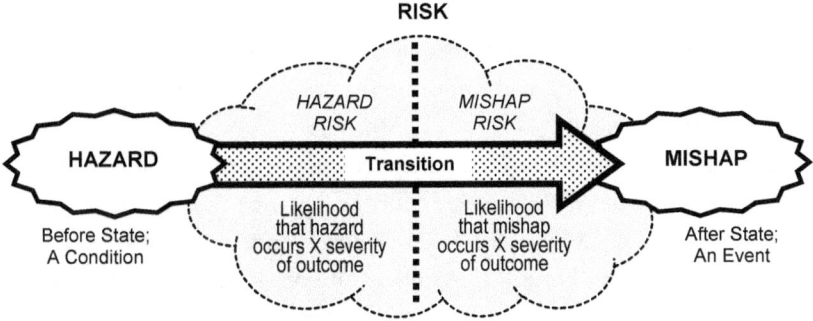

Figure 8.1 – Hazard Risk vs. Mishap Risk

Some basic axioms regarding safety risk include the following:

- Risk is a metric of the likelihood and consequence of a potential future event.
- When a hazard exists, there is always risk associated with it.
- Hazards and their components must be identified before risk can be assessed.
- Hazard risk varies based on the hazard components involved.
- Risk is in effect regardless of whether it's known or unknown.
- Risk can be eliminated, reduced or increased through design safety measures.

Risk is a measure that rates the relative safety significance (or danger) of a hazard. One thing is certain: hazards and risk exist regardless of our perception, knowledge or awareness of their presence. Hazards and risk do not care if we know about them or try to do anything about them. John Adams summed it up nicely: "Risk management: it's not rocket science – it's much more complicated"[8].

---

[8] John Adams, *Risk Management: It's Not Rocket Science – It's Much More Complicated,* Journal of System Safety, Mar-Apr, 2008, Pages 15-17.

## 8.2 Risk Rating

Measurement provides a mechanism for understanding an entity, and it also provides a means for evaluating changes to that entity. As Lord Kelvin stated "Anything that exists, exists in some quantity and can therefore be measured." Even though a hazard is a potential future event, its current value can be measured using the parameters of risk – likelihood and severity. Risk likelihood is the measure of the future event occurring, whereas risk severity is the measure of the amount of undesired consequence resulting from the future event when it occurs.

Two significant questions in the risk acceptance process (RAP) are, how should hazard risk be characterized for acceptance judgment and what acceptance criteria should be used. The risk acceptance method selected must address the concern of complexity versus utility. If the judgment criteria method is too complex it will not be used effectively. Figures 8.2 through 8.5 show four different example approaches.

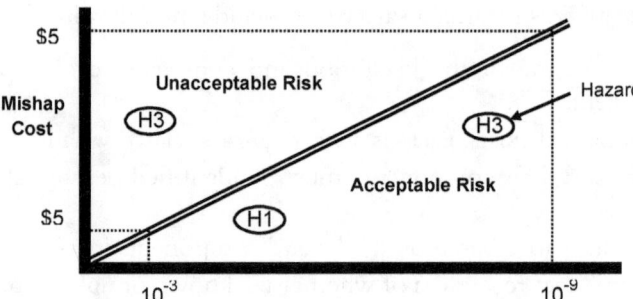

Figure 8.2 – Method 1: Above or Below Chart

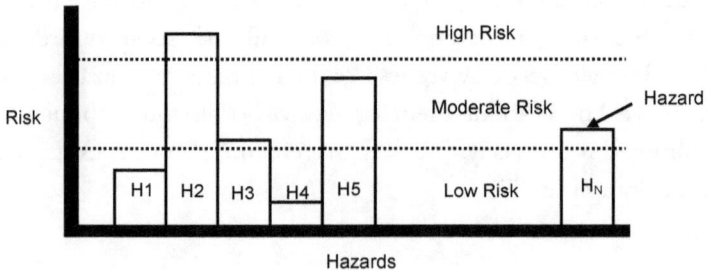

Figure 8.3 – Method 2: Histogram

| Probability | | Severity | | | |
|---|---|---|---|---|---|
| | | 1 | 2 | 3 | 4 |
| Frequent | A | 1 | 3 | 7 | 13 |
| Probable | B | 2 | 5 | 9 | 16 |
| Occasional | C | 4 | 6 | 11 | 18 |
| Remote | D | 8 | 10 | 14 | 19 |
| Improbable | E | 12 | 15 | 17 | 20 |

Figure 8.4 – Method 3: HRI Matrix

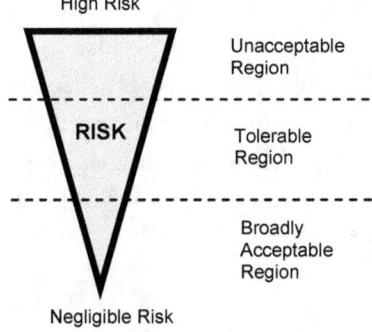

Figure 8.5 – Method 4: ALARP Chart

Method 1 shows an example of quantitative risk characterization by probability and severity cost. In this example, the mishap probability and cost of a hazard must be determined. The hazard risk point is then plotted on the graph and determined to be acceptable or unacceptable based on where it falls in the pre-defined acceptance region. In this method, mishap cost represents a method for measuring mishap severity. The major concern with this method is that there is little gradation in the acceptance level.

Method 2 shows an example of combined quantitative and qualitative risk characterization. In this example, all hazards are plotted by mishap risk on one axis. The hazards are then determined as acceptable or unacceptable based on where they fall in the pre-defined acceptance regions. The major concern with this method is how the risk metric is characterized (e.g., cost, deaths, etc.).

Method 3 shows a risk acceptance approach using the Hazard Risk Index (HRI) method suggested in MIL-STD-882. This method provides a good characterization of risk, which can be estimated qualitatively or quantitatively. It also provides a relatively simple methodology that is cost-effective to perform. The HRI approach from MIL-STD-882 is the most commonly used approach. Later versions of MIL-STD-882 refer to HRI as Mishap Risk Index (MRI), which is effectively the same thing. Appendix B describes this methodology in more detail.

Method 4 is known as ALARP, or As Low as Reasonably Practicable. If a given risk can be shown to have been reduced to as low a level as is reasonably practicable, taking into consideration the costs and benefits of reducing it further, then it is said to be a Tolerable Risk. The concept of diminishing risk proportion is shown by the triangle; as the risk is further reduced, the less proportionately it is necessary to spend to reduce it further. In the Unacceptable Region risk cannot be justified except in extraordinary circumstances. In the Tolerable Region risk is acceptable only if risk reduction is impracticable or if its cost is grossly disproportionate to the improvement gained. In the Broadly Acceptable Region risk is acceptable, but it is necessary to maintain assurance that the risk remains at this level.

# 9    HAZARD IDENTIFICATION

## 9.1 Hazard Identification

Hazard identification, evaluation and control constitute the backbone of the system safety process, which is achieved through a formal hazard analysis (HA) process. Understanding and thoroughly analyzing the system is a key aspect of effective HA. The HA process is performed on a specific system design configuration, since all of the aspects of that unique design must be considered in order to identify the system's unique hazards. If the design changes, the HA must change accordingly in order for the HA to correctly model the system. According to Chapanis[9], systems can be viewed from three different perspectives, each of which provides a different viewpoint and understanding of the system. These system viewpoints include:

1) Physical – the architectural view that depicts what the system contains and how it is constructed
2) Functional – describes what the system must do in order to produce the required system behavior, broken down into functions with input, output and transformation rules
3) Operational – defines how the user will view and operate the system, including instructions, conditions, parameters and limitations

It is important that HA consider and evaluate a system from each of these three perspectives in order to ensure complete safety coverage of the system. This is why HA must consider and focus on system functions, system operations, and system components, including hazardous energy sources.

---

[9] Chapanis, Alphonse, *Human Factors in Systems Engineering*, New York, Wiley, 1996.

This also explains why more than one type of HA must be applied in order to identify all hazards, because one HA type alone does not typically provide sufficient hazard identification coverage.

9.2 Hazard Analysis (HA)

Performing a good HA is not a simple or trivial task; it requires foresight, planning, methodology, organization and a total systems viewpoint in order to achieve uniformity, consistency and full system coverage. It also requires someone with experience and skill in safety and HA. Quite often analysts just jump into a system and immediately start identifying what they think are hazards, without considering the system architecture and the overall risk objectives and relationships for the HA. This approach often leads to confusion, overlap, gaps and hazard-risk mismatches.

The primary purpose of HA is to identify hazards and to obtain sufficient hazard data for risk assessments and design safety decisions. HA is applied to hardware, software, functions, procedures and human tasks. HA should be applied to the system at all stages of the system lifecycle. An HA becomes more detailed and accurate as more detailed information about the system becomes available. Different HA techniques and approaches are applied at different stages of the system life cycle to ensure all types of hazards are identified.

HA requires the systematic examination of a system, item or product within its lifecycle to identify hazardous conditions, including those associated with human, product and environmental interfaces, and to assess their consequences to the functional and safety characteristics of the system or product. In order to design-in safety, hazards must be designed-out (eliminated) or mitigated (reduced in risk), which can only be accomplished through HA. HA involves:

- Acquiring system data, knowledge and understanding
- Establishing HA scope and ground rules
- Analyzing the total system design to identify hazards and their causal factors
- Ensuring each hazard meets the criteria of the Hazard Triangle
- Determining the consequences resulting from an occurrence of the hazard
- Investigating any safeguards already in place to address the hazards
- Assessing the risk presented by the identified hazards

- Establishing design requirements to mitigate the risk presented by the hazards
- Formally documenting the entire process for safety case evidence

All of the basic system components and characteristics must be understood in order to perform a complete and thorough hazard analysis. Examples of typical safety considerations for various system elements include:

- Hardware – failure modes, hazardous energy sources
- Software – design errors, design incompatibilities
- Personnel – human error, human injury, human control interfaces
- Environment – weather, external equipment (e.g., radiation, chemicals)
- Procedures – instructions, tasks, warning notes
- Interfaces – erroneous input/output, unexpected complexities
- Functions – fails to perform, performs erroneously, performs inadvertently
- Facilities – construction factors, storage factors, transportation factors

An HA must adequately address each of these system attributes in order to ensure that all possible hazards are identified; all of these system elements and their interrelationships must be considered. For example, it is possible for different operational phases to have different safety impacts; different functions performed during a phase could have a direct impact on subsequent phases. During certain phases safety barriers or interlocks are often removed, making the system more susceptible to the occurrence of a hazard, e.g., at one point in the operational mission of a missile, the missile is powered and armed. This means that fewer potential failures are now necessary in order for a mishap to occur and there are fewer safeguards activated in place to prevent hazards from occurring.

Hazards are typically *actuated* as the result of any of the following system initiating mechanism factors, or combinations thereof:

- Hardware failures (aging, wear, random failure)
- Software errors/flaws (functional failure)
- Human errors (performance, decisions, judgment)
- Design errors/flaws (interface errors, sneak paths)
- External environmental factors (EMI, lightning)
- Maintenance flaws/errors (resulting in failures)
- Manufacturing flaws/errors (resulting in failures)

- Particular risk events (events occurring outside of a subsystem, such as a flood)

In addition to typical hazard sources, failures, flaws and errors can be also created or perpetuated by the following organizational factors:

- Poor design/development/manufacturing processes
- Organizational errors (performance, decisions, safety as a core value)
- Lack of safety culture in overall organization/company
- Safety organizational level of competence

It is important to note that hazards can exist and occur without the presence of a hardware failure mode or software error. Subtle design flaws can produce hazards, such as a sneak path in an electrical circuit. When performing an HA, standard considerations for identifying hazards include:

- Hazard Sources (e.g., energy sources, SCFs, environments, particular risk sources)
- Safety-related functions (SRFs) and safety-critical functions (SCFs) (hardware/software)
- Hardware failures (component failure modes)
- Sneak paths
- Safety-related (SR) and safety-critical (SC) human tasks

9.3 HA Complexity

A thorough, complete and logical HA is not a trivial process; it can be somewhat complex and difficult at times. There are several reasons for complexity in HA, such as:

- A typical hazard can be viewed in different ways by different analysts
- A typical hazard can be written in different ways by different analysts
- Hazards can sometimes be written to overlap due to carelessness and lack of planning
- Hazard risk levels can be different depending on the perspective and how the hazard is presented
- Risk can be manipulated through incorrect hazard identification and aggregation
- New technology combined in new ways without previous experience.
- System complexity
- System coupling (dependencies)
- System size

If an HA is not properly planned and performed it is easy to make errors and mistakes. Some common errors committed during the HA process that can significantly affect risk perception include the following:

- Writing a hazard at too high of a system level
- Writing a hazard at too low of a system level
- Writing a hazard incorrectly
- Failing to consider the interdependencies between certain hazards
- Attempting to combine multiple hazards into a single hazard
- Discarding a postulated hazard because it is assumed not plausible or credible

## 9.4 Credible vs. Non-Credible Hazard

Safety novices often assume that a non-credible hazard is one that has an extremely low likelihood of occurring and should therefore be discarded. This is not a valid assumption. A non-credible hazard is one that cannot feasibly happen (i.e., the Hazard Triangle cannot be sustained). A credible hazard is any hazard that can be postulated and which can feasibly happen, regardless of probability or likelihood. During HA, it is important to document postulated non-credible hazards and why they were determined to be such. This helps record for posterity that a complete HA was performed and all possibilities were considered.

## 9.5 Basic HA Approach

When performing an HA, the first requirement is a grounded understanding of the definition of a hazard; this involves understanding what a hazard comprises and how to correctly describe a hazard. The second essential element of HA is to utilize a specific and rigorous HA methodology. The third essential element of HA is to identify hazards at the correct structural and functional level of the system. If hazard risk is not assessed at the correct system level the risk can be understated, overstated, double counted or misinterpreted. The use of a system structural hierarchy establishes nomenclature that supports clear, unambiguous communication and definition of the system, its functions, components, operations, and associated processes. A system hierarchy refers to the organizational structure defining dominant and subordinate relationships between subsystems, down to the lowest component/piece part level. For proper risk assessment it is imperative that hazard risk be assessed at the proper system level, since it is very easy to write hazards at the inappropriate level.

There are many different HA techniques available, and a good SSP typically performs several different types of HA to ensure complete system hazard coverage. For any type of HA the recommended generic steps in the HA process include the following:

1) Obtain and understand the necessary system information:
   - System design and operation data
   - Design requirements, drawings, diagrams, etc.
   - System functions (hardware, software and user)
   - System components, limitations, environments, etc.
2) Develop HA support tools:
   - Table of system elements and their purpose/function (hierarchy list)
   - Table of system functions and their purpose
   - Table of hazard sources used in the system
   - Table of particular risk sources applicable to the system
   - Table of expected environments
3) Plan the HA process and methods before starting any HA:
   - Establish the overall approach, coverage, methodologies and ground rules
   - Establish the level of analysis
   - Establish the method of documentation (evidence)
4) Perform the actual HA:
   - Analyze functions (hardware and software) for hazards if the function should fail, occur inadvertently or function erroneously
   - Analyze components, assemblies and subsystems for hazards should these items fail, occur inadvertently or function erroneously
   - Analyze hazardous energy sources for hazards should these items fail, occur inadvertently or function erroneously
   - Analyze all human performed tasks for hazards if the task should fail, be performed out of sequence or be performed erroneously
   - Identify existing safety features and re-engineer the hazards that established the need for these features
   - Analyze design requirements for hazards should these requirements fail, occur inadvertently or function erroneously
5) Use the following keywords for each item analyzed, as an aid in hazard recognition:
   - Fails to operate

- Operates incorrectly/erroneously
- Operates prematurely/inadvertently
- Operates out of sequence
- Operates in degraded mode
- Unable to stop operation as intended
- Sends/receives/displays erroneous data
- Causes operator confusion

6) Write hazards to ensure they are characterized in complete system context:

- Describe the Hazard Source, Mechanism and Target/Threat Outcome
- Do not abbreviate or assume reader understands program-special lingo
- Describe the hazard scenario in context (e.g., "fuel" is not a hazard, but, "fuel leak and an ignition source leading to fire and system loss" is a hazard)
- Make sure the hazard context includes the specific hazard causal factors and the specific effects

7) Continuously review and compare results to ensure the analyses are correct and thorough

## 9.6 Minimum HA Coverage

How many HAs should be performed on a typical system? Typically, multiple HA methodologies are applied to a system to ensure complete system coverage and complete hazard identification. This is necessary to ensure complete safety coverage of system complexity and to cover the three system viewpoints mentioned earlier: physical, functional and operational. This is not a hard and fast rule, but depends upon various factors, such as: system size, system complexity, safety criticality of the system, etc. The number and type of HAs to be performed should be planned in advance and tailored to the specific system and the system requirements involved.

Over 100 different HA techniques have been developed over the years. However, many of these techniques are not true HAs and many are merely variations of other HA techniques. There are only about 15 to 20 HA techniques that are commonly used by system safety experts. The primary HA techniques recommended by system safety experts and best practice standards are identified in chapter 4. As already stated, tailoring is recommended, but as a minimum the following HA techniques should always be performed:

- Preliminary Hazard List (PHL) analysis
- Functional Hazard Analysis (FHA)
- Preliminary Hazard Analysis (PHA)
- System Hazard Analysis (SHA)

## 9.6.1 Preliminary Hazard List (PHL) Analysis

The PHL is not a complete formal HA, but is a very preliminary high level analysis done as a precursor to more detailed HAs which follow. The original intent of the PHL was to roughly identify hazards during the conceptual design phase, which would later be expanded upon by more detailed HA methods. Over the years the objective of the PHL has changed slightly and it now typically provides hazard sources and hazard categories. Typical PHL information includes:

- High level hazards (not yet detailed)
- Hazardous components and functions in the system
- Safety-related (SR) and safety-critical (SC) functions
- Top Level Mishap (TLM) and Top level hazard (TLH) categories
- System vulnerabilities – zonal, particular risk and environmental risk sources

The PHL identifies basic hazards, hazard sources and hazard initiating mechanisms, and forms the initial Top Level Mishap (TLM) categories. The TLM categories are used to classify hazards into common categories which help to establish a hazard relationship diagram. The more detailed HA methodologies use information from the PHL to identify and write the actual hazards.

## 9.6.2 Functional Hazard Analysis (FHA)

Typically, system functions are defined and developed before system hardware is designed. For this reason it is important to assess the functions as early as possible in the design program in order to quickly start identifying hazards. The FHA evaluates system functions to identify functional hazards and to determine which functions are SR and SC. Functional hazards establish a functional hazard thread, whereby all of the hardware and software in that thread are potential hazard sources and/or initiating mechanisms. Functions and functional hazard threads that are tagged as SR and SC will receive more scrutiny in the HAs that follow. The FHA provides the following HA information: functional hazards, functions that are SR and SC, and equipment and components, in the functional hazard threads, that may present hazards, hazard sources and initiating mechanisms

### 9.6.3 Preliminary Hazard Analysis (PHA)

The PHA is probably the most commonly performed hazard analysis technique. In most cases, the PHA identifies the majority of the system hazards. The remaining hazards are usually uncovered when subsequent hazard analyses are generated and more design details are available. Subsequent hazard analyses refine the hazard cause-effect relationship, uncover previously unidentified hazards and refine the design safety requirements.

Use of the PHA technique is highly recommended for every program, regardless of size or cost, to support the goal of identifying and mitigating all system hazards early in the program. The PHA is the starting point for further hazard analysis and safety tasks, is easily performed, and identifies a majority of the system hazards. The PHA is a primary system safety hazard analysis technique for an SSP.

The purpose of the PHA is to identify hazards, hazard causal factors, hazard mishap risk and SSRs in order to mitigate hazards with unacceptable risk. The PHA provides the following HA information: hazards, mishaps, hazard causal sources, hazard risk, SCFs, TLMs, Hazard mitigation methods and recommended mitigation methods.

### 9.6.4 System Hazard Analysis (SHA)

The SHA is an analysis methodology for identifying hazards and evaluating risk and safety compliance at the system level, with a focus on interfaces, Safety Critical Functions (SCFs) and particular risk events. The SHA ensures that identified hazards are understood at the system level, that all causal factors are identified and mitigated, and that the overall system risk is known and accepted. SHA also provides a mechanism for identifying previously unforeseen interface hazards and evaluating causal factors in greater depth.

The SHA is a detailed study of hazards resulting from system integration. This means evaluating all identified hazards and hazard causal factors across subsystem interfaces. The SHA expands upon the PHA and the Subsystem Hazard Analysis (SSHA) and may use additional techniques, such as Fault Tree Analysis (FTA), to assess the impact of certain hazards at the system level. The system level evaluation should include analysis of all possible causal factors from sources such as design errors, hardware failures, human errors, software errors, etc. The SHA provides the following HA information: hazards, mishaps, hazard causal sources, hazard risk, SCFs, TLMs, Hazard mitigation methods and recommended mitigation methods.

The SHA assesses the safety of the total system design by evaluating the integrated system. The primary emphasis of the SHA, inclusive of hardware, software, and Human Systems Integration (HSI), is to verify that the product is in compliance with the specified and derived system safety requirements (SSRs) at the system level. This includes compliance with acceptable mishap risk levels. The SHA examines the entire system as a whole by integrating the essential outputs from the SSHAs. Emphasis is placed on the interactions and the interfaces of all the subsystems as they operate together.

An analysis of *particular* risk sources is part of the SHA. Particular risk refers to risk associated with those hazard sources or influences that are outside the subsystem of interest, but which may violate failure independence claims within the system. A particular risk analysis examines those particular risk events or influences that have the capacity to adversely affect the system redundancy and design diversity. Particular risk sources are essentially common cause failure events. For example, a system fire could simultaneously cause loss of several redundant items, even though they are physically separated. These particular risks may also influence several zones (e.g., aircraft) at the same time. Some of these risks may also be the subject of specific aircraft airworthiness requirements. Some of the known particular risk sources result from airworthiness regulations, while others arise from known external threats to aircraft or other system types. Typical particular risk sources include, but are not limited to, the following:

- Fire
- High energy devices (engines, motors, fans)
- Leaking fluids (fuel, hydraulic, battery acid, water)
- Hail, ice, snow
- Bird strike
- Tread separation from tire; wheel rim release
- Lightning
- High intensity radiated fields
- Flailing shafts
- Bulkhead rupture (e.g., aircraft, ship)

# 10    SYSTEM SAFETY TOOLS

## 10.1 Introduction

As previously stated, hazard identification, evaluation and control constitute the backbone of the system safety process. The only sure way to thoroughly identify hazards is through the application of rigorous, structured and formal hazard analysis (HA) techniques. A formal HA technique is necessary to provide analysis rigor and also to provide analysis evidence. There are many different tools available to the system safety analyst for hazard recognition, recording, evaluation and tracking. The tools in the safety toolbox essentially fall into the following categories:

- Hazard identification tools
- HA support tools
- Hazard recognition aids
- Historical data
- Hazard tracking tools

## 10.2 Hazard Identification Tools

Typically, multiple HA methodologies are applied to a system to ensure complete system coverage and complete hazard identification. This is necessary to provide complete safety coverage of the three system viewpoints mentioned earlier: physical, functional and operational. The number and type of HAs to be performed should be planned in advance and tailored to the specific system and the system requirements involved, depending upon various factors, such as: system size, system complexity, safety criticality of the system, etc.

Within the system safety discipline there are over 100 different HA techniques that have been proposed, some of which are unique, some of which are variants of others, some of which are extremely useful and some of which arc not useful at all. Some of the techniques are not true HAs, and many are merely variations of other HA techniques. There are only about 15 to 20 HA techniques that are commonly used by system safety experts. Each HA technique has a unique format methodology. Essentially, there are *primary* and *secondary* HA techniques. The primary HA techniques are full-fledged, or complete, methodologies for identifying all system hazards. The secondary HA techniques typically do not identify all hazards and are support for the primary techniques; many help identify the root causal factors of already identified hazards. Some of the most important and most used HA techniques include:

A) Primary HA techniques
- Preliminary Hazard List (PHL) analysis
- Preliminary Hazard Analysis (PHA)
- Subsystem Hazard Analysis (SSHA)
- System Hazard Analysis (SHA)
- Operations and Support Hazard Analysis (O&SHA)
- Health Hazard Assessment (HHA)
- Threat Hazard Analysis (THA)

B) Secondary HA techniques
- Fault Tree Analysis (FTA)
- Event Tree Analysis ETA)
- Bent Pin Analysis (BPA)
- Sneak Circuit Analysis (SCA)
- Safety Requirements/Criteria Analysis (SRCA)
- Barrier Analysis
- Interlock Analysis
- Failure Mode and Effects Analysis (FMEA)
- Code Safety Analysis
- Particular Risk Analysis
- Markov Analysis (MA)

It should be noted that many analysis techniques are incorrectly used for HA, such as Failure Mode and Effects Analysis (FMEA). FMEA results can be used as input information for an HA, but the FMEA does not suffice for an HA because it does not thoroughly cover system hazard-mishap scenarios and it does not cover the combined effect of multiple simultaneous failures that often cause hazards. An FMEA is not a hazard analysis or a safety analysis and should not be used in place of either, but it can be used to supplement them as a secondary supporting technique.

MIL-STD882 recommends that the seven primary HA techniques of: PHL, PHA, SSHA, SHA, O&SH, HHA and SRCA always be performed on each system development program. Back in chapter 4, Figure 4.3 contains a sample milestone chart showing the time period when each of these techniques is typically performed during the system lifecycle. Depending upon the size, complexity and criticality of the system, the specific HA techniques to be applied can be tailored to the program. See "Hazard Analysis Techniques for System Safety" by Clifton A. Ericson II, July 2005, Wiley, for a detailed description of 24 of the most used HA techniques and quasi-HA techniques.

10.3 HA Support Tools

HA is a complex and thoughtful process, which can be simplified with the help of several design development tools that are usually available on a development program. The tools most useful in HA are typically produced by the systems engineering and reliability engineering disciplines.

10.3.1 System Engineering Tools

A system can be very large, comprising hundreds of subsystems, thousands of assemblies and millions of components. The systems engineering process utilizes many different tools during systems development that aid in the understanding of large and complex systems. Some systems engineering tools that greatly aid the system safety analyst are:

- Simplified System Diagrams
- Functional Block Diagrams
- Master Equipment List (MEL)

Simplified system diagrams assist the safety analyst in understanding how a system operates by viewing the subsystems and components within a system, along with their interface points. These diagrams help the safety analyst identify system hazards, interface hazards and functional hazards. They also help identify hazard sources in the system.

The Functional Block Diagram (FBD) is the systems engineering tool for showing how a system operates, by depicting components, functions and interfaces in a single diagram. The FBD is also known as the Single Line Diagram (SLD) or the Functional Flow Diagram (FFD). The FBD provides a means whereby individuals not familiar with the detailed design can quickly and easily grasp how the system operates. For example, an FBD could be used to translate a complicated electric circuit diagram into a simplified series of functions.

The Master Equipment List (MEL) is the systems engineering tool for exhibiting both system components and system hierarchy. Basically, the MEL is a list of all the systems, subsystems, units, assemblies and components in the major system, with each item in the list indented to reflect its hierarchy and ownership level. The MEL is a system hierarchy table.

10.3.2 Reliability Engineering Tools

Some reliability engineering tools that greatly aid the system safety analyst include the following:

- Reliability Block Diagrams (RBDs)
- Failure Mode and Effects Analysis (FMEA)

The Reliability Block Diagram (RBD) is a reliability engineering tool for showing how system elements are logically tied together for purposes of reliability evaluations. These diagrams depict series systems, parallel systems and complex combinations of the two. These diagrams are also a key to how a system functions.

Failure Mode and Effects Analysis (FMEA) is a single-thread analysis of all the components in an item and considers all the potential failure modes of the components in that item. The failure modes are then evaluated for the effect of the failure mode and the failure rate of the failure mode. An item can be an assembly, sub-subsystem, process or even the system, depending upon the level of system detail under investigation. FMEAs provide the safety analyst with failure modes that might cause or contribute to hazards, and it provides failure rates for safety risk assessments. It must be stressed that an FMEA provides useful safety information, but does not suffice for an HA.

10.4 Hazard Recognition Aids

Hazard recognition is the cognitive process of visualizing a hazard from an assorted array of design information. Hazard recognition and identification is not always an easy process. Hazards are like ubiquitous yet elusive little creatures that must be hunted and captured, where the hunt is akin to the HA process. There are, however, some aids which assist the analyst in the process, such as:

- Hazard Checklists
- Mishap Checklists
- Hazard Recognition Keywords
- Historical Data

## 10.4.1 Hazard Checklists

Hazard checklists are existing checklists of known hazard sources. Hazard checklists focus on the Hazard Source element of the Hazard Triangle. There are many different types of hazard checklists, such as energy source checklists, hazardous operations checklists, hazardous environment checklists, etc. This means that generic hazard source checklists can be utilized to help recognize hazards within the unique system design. If a component in the system being analyzed is on one of the hazard checklists, then this is a direct pointer to potential hazards that may be in the system. It is important to postulate all of the possible ways that the checklist item can be hazardous within the system design. It is incumbent on the system safety analyst to obtain, maintain and utilize as many checklists as possible.

## 10.4.2 Mishap Checklists

Hazards can be recognized by focusing on known or past undesired outcomes or mishaps. Mishap checklists focus on the Target/Threat Outcome element of the Hazard Triangle. This means considering and evaluating known undesired outcomes that are applicable to the system under analysis. For example, a missile system has certain undesired outcomes that are known right from the conceptual stage of system development. By following these undesired outcomes backward, certain hazards can be more readily recognized. In the design of missile systems, for example, it is well accepted that inadvertent missile launch is an undesired mishap, and therefore any conditions contributing to this event would formulate a hazard, such as auto-ignition, switch failures and human error.

### 10.4.3 Hazard Recognition Keywords

Another method for recognizing hazards is through the use of hazard recognition keywords. This is a method involving a set of clue questions that must be answered, each of which can trigger the recognition of a hazard. The keywords are potential states or ways the subsystem could fail or operate incorrectly and thereby result in creating a hazard. For example, when evaluating each subsystem, answering the question "what happens when the subsystem fails to operate?" may lead to the recognition of a hazard. Each system safety analyst should develop their specialized and unique list of hazard recognition keywords. Some example keywords include the following:

- Fails to function (operate)
- Functions incorrectly/erroneously
- Functions prematurely/inadvertently
- Functions erroneously (out of sequence, degraded mode)
- Unable to stop function as intended
- Sends/receives/displays erroneous data
- Causes operator confusion

### 10.4.4 Historical Data

Another method for recognizing hazards is through the use of past knowledge from experience and lessons learned. Safety lessons learned from a previously developed system that is similar in nature or design to the system currently under analysis will aid in the hazard recognition process. For example, by reviewing the lessons from an earlier spacecraft system, it might be discovered that the use of a particular seal design on a specific part of the spacecraft resulted in several mishaps. Seal design on this part of the spacecraft should then be recognized as a potential hazardous area requiring special focus in current new spacecraft design. Historical data is a valuable asset that can aid in the identification of hazards. Some common sources of useful information include the following:

- Mishap/accident/ reports
- Incident reports
- Lessons learned databases
- Customer complaints
- Maintenance and repair records

Much can be learned from historical failure data, past mistakes and lessons learned. We are doomed to repeat past mistakes if we do not keep a record of them and continuously study and apply them in HA. However, there must be an established documented protocol and procedure to track hazards in order to use said information in the evaluations of incidents.

## 10.5 Hazard Tracking Tools

Hazard tracking is the process of systematically recording all identified hazards and the data associated with these hazards as they progress through the lifecycle of a hazard: identification, assessment, mitigation, verification, acceptance and closure. It is typically achieved through the use of a formal electronic hazard tracking system (HTS). An effective HTS actually establishes a process that facilitates record keeping, generating reports and maintaining the necessary evidence of a safety program. A well designed HTS is an effective tool for the SSP.

# 11    SYSTEM SAFETY VALUES, AXIOMS AND PRINCIPLES

## 11.1 System Safety Values

System safety values are basic values that guide the system safety discipline and system safety practitioners. The following are the core values of the system safety discipline:

Value # 1: People

System safety is about saving lives; it involves protecting people from hazards and mishaps. This is achieved by applying the system safety process to design and develop systems that are not going to kill, injure or maim people through mishaps.

Value # 2: Prevention

System safety is about preventing mishaps before they kill or injure people. Potential mishaps are foreseeable and preventable. Prevention is more effective than correction; it saves lives and money. This is achieved through the system safety process of identifying and eliminating/mitigating hazards.

Value # 3: Integrity

System safety is about doing the right things to protect people from mishaps and hazards. This is achieved by putting safety first in the design and development of systems. The system safety process is built around the integrity of a lifecycle and closed-loop process that ensures safety is built directly into the system design. No hazard should be accepted when that hazard can be reasonably eliminated or reduced in risk via design safety measures.

Value # 4: Responsibility

System safety is about being responsible for a safe system design. The slogan that "safety is everyone's responsibility" works well in a shop or office environment, but developing a safe system design requires a group solely responsible for, and focused on, design safety. This is achieved by dedicated group of system safety specialists experienced in design safety. Only a disciplined system safety program can provide the focus and experience needed to cover all the boundaries and interfaces involved in a system.

Value # 5: Process

System safety is about implementing and following a scientific and engineering process that has been proven to develop safe system designs. This is achieved using the system safety process which has a proven, formal, systematic and systems oriented methodology for the identification and control of hazards and risk. System safety applies a detailed system focus on design, hazards and risk across the entire lifecycle of a system, including all interfaces, environments and operational tasks.

Value # 6: Design

System safety is about building safety directly into the system design. When the actual system design is safe, safety does not have to be added in later by corrections or through the use of special procedures. This is achieved by the design-for-safety philosophy of the system safety process. Because of the complexity and uniqueness of systems, safety cannot be totally achieved through a prescriptive compliance based process, true safety requires a design-for-safety risk based approach; a holistic "systems" approach.

Value # 7: Business

System safety is about following a business case that saves money in the long run; safety is a strategic business element. This is achieved by applying the system safety process, which prevents mishaps. It's cost effective to initially anticipate and control hazards, rather than fixing systems after mishaps occur, and paying the mishap costs, design fix costs and intangibles costs that are incurred. Safety is not free, but it is typically cheaper than the alternative.

11.2 System Safety Axioms

System safety axioms form basic truths, laws or tenets that are used as a basis for system safety reasoning. These axioms articulate the basic truisms of the system safety process and discipline; they form the foundation for developing safe system designs. The following are the core axioms of the system safety discipline:

Axiom #1:  Systems are essential to human activity and human life.
- We live in a world involving systems and systems-of-systems.
- We interface and use systems daily.
- Systems can help us or hurt us; systems have no conscience.
- Systems provide a utility, but also impose a continual conflict between man and machinery.

Axiom #2:  Systems contain the components that cause mishaps.
- Every system contains inherent potential safety vulnerability.
- Safety vulnerability is characterized in terms of hazards.
- Most systems typically contain one or more hazard sources, which spawn safety vulnerability and potential mishaps.
- Every system containing a hazard source has a unique set of hazards associated with that source.
- Some systems have more vulnerability than others, depending on system size, function, safety criticality and hazard sources involved.

Axiom #3: Mishaps are foreseeable, predictable and controllable.
- Mishaps are undesired events resulting from the occurrence of specific causal factors.
- Mishaps don't just happen indiscriminately; they are not random chance events.
- Hazards and mishaps are directly linked; hazards are the precursor to mishaps.
- Potential mishaps are characterized by hazards.
- System hazards are man-made; they are the result of specific design flaws.

<u>Axiom #4</u>: Hazard identification and elimination is the key to system safety.

- A hazard defines a potential mishap and its associated potential mishap risk.
- Hazards have three states: dormant, armed and active.
- All hazards are comprised of three basic required constituent components (S, M, TTO).
- Hazards can be recognized using the S-M-TTO hazard model.
- Hazards (and risk) can be controlled by modification of any of the three hazard components.
- System safety hazard analysis provides the means for hazard identification and control.
- Most hazards attributed to human error are basically design-safety errors, where the true causal factors are discovered via system safety hazard analysis.

<u>Axiom #5</u>:  Hazards are best eliminated/mitigated using the system safety process.

- Since mishaps are defined by hazards, then safety is not achieved unless all hazards are identified and controlled.
- Safety is optimally achieved when hazards are proactively identified and eliminated or mitigated (i.e., before a mishap, not after).
- The system safety process focuses predominantly on hazard identification and elimination/mitigation.
- The system safety process utilizes the incorporation of design safety methods, features and/or appliances into the system design in order to eliminate/mitigate hazards.
- Hazard identification and elimination must be initiated at the beginning of system design, with safety features intentionally designed into the system.
- System safety follows a Safety Order of Precedence for ideal hazard mitigation and optimal safety: 1) design methods, 2) safety devices, 3) warning devices, and 4) procedures/training.
- Hazards with catastrophic outcomes should not be mitigated solely through the use of safety procedures that depend upon human decisions and actions.

<u>Axiom #6</u>:  Risk is a safety tool for precision hazard elimination/mitigation.

- The amount of danger a hazard presents is characterized by the metric of risk.
- Risk is a measure of hazard occurrence likelihood combined with hazard outcome severity.
- Many hazards cannot be eliminated because the hazards sources must remain in the system, thus the risk must be reduced.
- Risk management is not an excuse to avoid eliminating/mitigating hazards; it is a tool for measuring the amount of risk reduction.
- No hazard should be accepted when that hazard can be reasonably eliminated or reduced in risk via design safety measures.

<u>Axiom #7</u>:  The system safety process can be applied to any product, system or process.

- System safety is a proven rigorous, disciplined and systematic methodology that applies critical thinking.
- System safety focuses on hazard elimination via system design-safety.
- System safety employs a total systems approach, as opposed to a piece-part approach, a behavior based approach or a minimum standards approach.
- System safety has universally been successfully applied to a diversity of products, systems and processes.

<u>Axiom #8</u>:  The system safety process is cost effective.

- System safety prevents mishaps by eliminating/mitigating hazards.
- Prevented mishaps save tangible resources – lives, physical assets, time and monetary.
- Prevented mishaps prevent intangible losses – reputation, potential sales, etc.
- Prevented mishaps prevent liability lawsuits.

11.3 System Safety Principles

System safety principles are the fundamentals beliefs that define the safety ideal. These principles are not "lip service," a "slick slogan" or a plaque placed on a wall. These fundamentals are key values that are important to a successful and ethical system safety program (SSP) and system safety process; they form the safety foundation for the development of safe systems. These fundamental principles should always be applied by an SSP, and they can be translated into safety design requirements as necessary. Standard system safety principles are like proverbs, which include the following:

- All mishaps are preventable; system safety is the process for achieving this via the elimination and mitigation of hazards.

- Hazard identification, hazard elimination/mitigation and risk reduction require a documented system safety process with specified artifacts as evidence of completion and success.

- Safety must be shown to exist in the system design rather than assuming it exists. This is primarily achieved through the core system safety process and the resulting artifacts, which together establish a safety case for the system or product.

- A system must be proven to be safe, not just assumed to be safe. This requires the implementation of an SSP, implementing the core system safety process that develops a safety case for the system.

- In the case of safety, the system should be considered guilty (unsafe) until proven innocent (safe).

- Mitigation verification requires *evidence* in the form of testing; analysis and/or inspection is allowed when testing cannot provide the evidence.

- Safety must be a core value of the company in order for safety to compete with other factors, such as cost and schedule. A safety culture is difficult to establish and maintain if safety is not a core value. If safety is made merely a *priority*, it should be realized that priorities often change under stressful conditions and are sacrificed for cost and schedule, whereas core values are not downgraded.

- Everyone on the engineering staff should receive specialized training in system safety by experienced and knowledgeable system safety specialists. This will help designers in the design safety process.

- A system safety program (SSP) must have a manager and staff specifically qualified and experienced in system safety.

- If human factors and potential human errors are omitted from hazard analyses, then the safety case is incomplete and system risk understated.

- In order to identify all potential hazards and have a complete and thorough hazard analysis, software and human factors must be analyzed as an integral part of the system.

- Only a formal hazard analysis can effectively identify hazards and hazard-mishap risk, and only a qualified system safety practitioner should perform Hazard Analyses.

- Failure Mode and Effects Analysis (FMEA) provides an aid to hazard analysis, but it does not suffice for a hazard analysis or safety analysis.

- The cost of a safety program varies by system size, complexity, and criticality. How much to spend on a safety program can be judged by how much all of the total losses would cost if all the major potential mishaps in the system occurred. Mishaps costs can be estimated by evaluating all of the possible worst-case mishap scenarios and summing the outcome costs.

- Not all hardware failures, software bugs or human errors are of safety concern, only those pertaining to hazards and mishaps.

- The company developing a system/product must ensure their subcontractors and suppliers properly support the SSP, including plans, tasks and data.

- Commercial-off-the-shelf (COTS) items must be evaluated for system safety as an integral part of the entire system. COTS items are *not* exempt from the system safety process or from safety requirements of the system.

- In consumer products and systems, safety-critical safety warnings should never be placed solely in an operator's manual or user's manual, because the operator or user will likely never read the manual. The system should be designed to warn the user. Possible methods include sound, labels, displays, etc.

- Humans will naturally err, but most mishaps attributed to human error are actually design errors, which allow or force the operator to commit a safety-critical error[10]. The system should be designed to prevent or counter all potential human errors that could contribute to a safety-critical mishap. Specifically the system design should:
  - Not allow the system to injure the operator
  - Not confuse the operator into committing an error
  - Be resistant to potential human errors that are safety-related
  - Counter potential safety-critical errors that might be performed

---

[10] Stems from paper by Dr. Alphonse Chapanis titled "To Err Is Human, To Forgive, Design" presented at the 25th annual ASSE professional development conference and exposition in New Orleans, 1986.

- To err is human, but to allow errors to kill is not acceptable and is preventable (through the combined efforts of system safety and human factors engineering).

- Mishaps attributed to human error may actually be due to design errors if the causal factors are carried back far enough. Quite often the system design contributes to a human error, or even forces the human to err.

- The system design shall contain safety design features (DSFs) to minimize mishap risk during all life cycles phases of the system.

- No hazard should be accepted when that hazard can be reasonably eliminated or reduced in risk via design safety measures.

- Accept risk only when benefits outweigh the potential mishap damages, losses and costs.

- Risk information, risk knowledge and risk management is not an excuse for not eliminating or mitigating hazards.

- Hazards with catastrophic outcomes should not be mitigated solely through the use of safety procedures that depend upon human decisions and actions.

- Merely establishing warnings and instructions is no substitute for performing system safety engineering.

# 12   DESIGN SAFETY METHODS

## 12.1 Design Safety Feature

System design, or architecture, is the mechanism that establishes the safety level of a system. A Design Safety Feature (DSF) is a special feature, or mechanism, intentionally placed in the system design for the purpose of mitigating a specific hazard and its associated risk. It is incorporated into the design of a system or product specifically to improve the safety attribute of a system. A DSF may not be necessary for system function, but it is necessary for hazard elimination or mitigation. A DSF can be any device, technique, method or procedure incorporated into the design to specifically eliminate or reduce the risk factors forming the hazard.

In order to effectively eliminate or mitigate a hazard, it is necessary to fully understand the causal factors involved and then selecting DSFs that can counter these factors and reduce the risk. DSFs should be chosen based on effectiveness, cost, and feasibility for each specific hazard. Feasibility includes consideration of both means and schedule for accomplishment. During product and system development design reviews the DSFs should be identified to show that an effective system safety program (SSP) is in effect. This is also a time to identify and take credit for DSFs in the product/system design.

It should be noted that the terms DSF, Safety Device, Safety Feature, Safety Measure, Safety Mechanism, Design Safety Measure, Design Safety Mechanism and Hazard Countermeasure are synonymous and can be used interchangeably; however, most system safety practitioners tend to use the term DSF most often.

12.2 Hazard Risk Reduction

There are two ways to mitigate the risk presented by a hazard: a) reduce the probability of the hazard occurring and b) reduce the severity of the hazard-mishap outcome, as depicted by Figure 12.1. In reality, it is usually very difficult, if not impossible, to change the potential severity outcome of most hazards. Therefore, the primary way to reduce risk is to attack and counter the hazards causal factors and reduce their likelihood of occurrence.

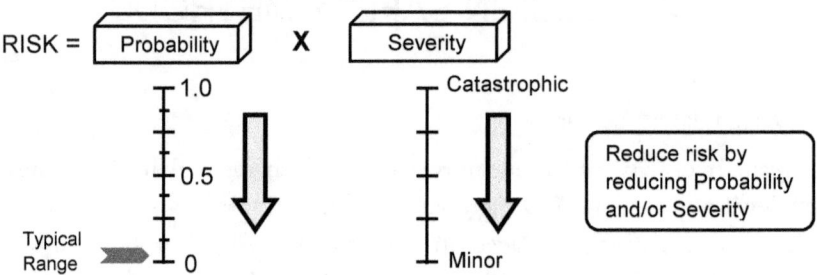

Figure 12.1 – System Safety Mishap Risk Goals

12.3 Safety Order of Precedence (SOOP)

When developing DSF options to mitigate hazards, it is recommended that the Safety Order of Precedence (SOOP) be followed. This safety hierarchy protocol was originally established by MIL-STD-882. The SOOP provides a preferred order for implementing different DSFs, each of which affects the level of risk differently; each SOOP level provides a higher level of safety assurance in a step-like manner. Figure 12.2 graphically displays the Safety Order of Precedence (SOOP). The thicker part of the bar indicates the strength and desirability of the safety feature.

It should be noted that hazard mitigation does not necessarily have to be limited solely to one of the options in the SOOP; incorporating one or more of the SOOP options may mitigate a hazard. However, the best and recommended option is to always attempt to reduce mishap risk through design measures, and then utilize other measures as necessary.

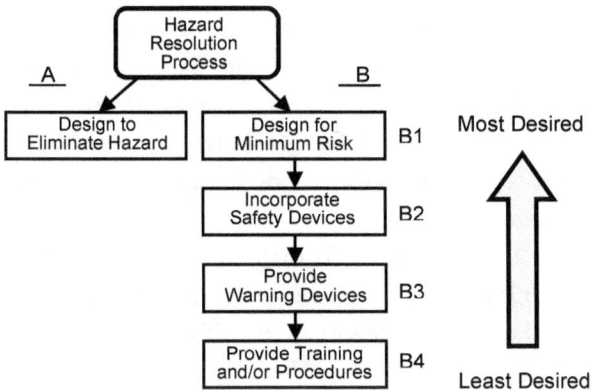

Figure 12.2 – Safety Order of Precedence

When the design development effort has been advanced beyond basic concepts, designers will often request that a hazard be controlled by procedural instructions and warning in order to avoid costly design changes. Managers should recognize these safety features are the lowest preference remedies. These remedies depend upon human action, or a lack of such action, to prevent the hazard from developing into a mishap. System designers should therefore resist attempts to substitute human action for good design. The use of man-dependent controls for Category I hazards should be expressly prohibited since the required levels of probability cannot be met (e. g., human error rates vary from $1 \times 10^{-1}$ to $1 \times 10^{-4}$, while minimum acceptable risk for Category IE is $1 \times 10^{-6}$). Intended human performance is the least dependable means of averting an accident; it requires knowledge and training, attentiveness, willingness to exert the necessary effort, even under conditions of fatigue, and frequently requires dependable lighting to permit observation of a warning sign.

## 12.4 Typical Design Safety Features

The purpose of a DSF is to mitigate a hazard and its associated risk. A hazard mitigator is a specific design method applied to a specific identified hazard. Sometimes there is confusion in regard to methods of mitigation; what some people consider to be a hazard mitigation method may not actually be one. For example, high integrity development processes do not mitigate hazards, they merely provide a level of assurance that, maybe, no hazard

# C. A. Ericson II

exists due to the stringent level of development rigor applied. It should be noted that not all types of testing are actual hazard mitigations. For example, black box testing, white box testing and requirements testing verify that implemented mitigation methods are working. Testing also shows, to some degree, that no unknown hazards exist. To mitigate a hazard, the hazard must be identified and then the causal factors must be specifically targeted for mitigation. A general rule to remember: one hazard, one or more specific and direct DSFs to mitigate the hazard.

DSFs can take many different forms and methods. Table 12.1 contains an example list of different methods and techniques that are commonly used to enhance safety. The SOOP category is taken from Figure 12.2.

Table 12.1 – Typical DSF Methods

| DSF | Category | SOOP Category |
|---|---|---|
| Redundancy design | B1 | Design |
| Fail-safe design | B1 | Design |
| Fault-tolerant design | B1 | Design |
| No single-point failure design | B1 | Design |
| Design diversity | B1 | Design |
| Barriers | B1 | Design |
| Partitions | B1 | Design |
| Interlock (inhibit) design | B1 | Design |
| Intrinsic Safety design | B1 | Design |
| Isolation design | B1 | Design |
| Enhanced part reliability | B1 | Design |
| Weak link design | B1 | Design |
| Safety factors | B1 | Design |
| Safety margins | B1 | Design |
| Fail-operational design | B1 | Design |
| Fail-non-operational design | B1 | Design |
| Energy control design | B1 | Design |
| Fire detection and suppression | B2 | Safety Device |
| Personal protective equipment (PPE) | B2 | Safety Device |
| Survival system design | B2 | Safety Device |
| Lockout/Tagout | B2 | Safety Device |
| Warnings and Cautions | B3 | Warning Device |
| Special procedures | B4 | Procedure |
| Special training | B4 | Training |
| Scheduled X-ray analysis before use | B4 | Procedure |
| Scheduled maintenance before use | B4 | Procedure |
| Scheduled inspection before use | B4 | Procedure |

In general, safety devices often tend to be static interveners intended to serve as hazard countermeasures. Examples include physical guards, barricades, revetments around explosive storage facilities, guardrails, machine guards, safety eyewear, hearing protection, guards, and barricades.

## 12.5 Hazard Mitigation Example Using SOOP

To illustrate the safety order of precedence, an exaggerated example is shown in Table 12.2. In this example, the landing gear for a commercial aircraft has been analyzed for hazards, and the following hazard has been identified:

> Hazard: "Aircraft retractable landing gear fails to extend for landing, resulting in a gear-up crash landing that causes death/injury of passengers."

Table 12.2 – Proposed Landing Gear Hazard Mitigations

| Precedence Order | Method of Resolution |
|---|---|
| Eliminate Hazard | Use fixed landing gear (i.e., always extended down). |
| Reduce Hazard | Provide redundant extension systems with redundant power and controls. |
| Safety Device/Feature | 1) Provide manual hand crank as a backup safety device. <br> 2) Perform a self-test of the system prior to retracting; prohibit retraction if failure is discovered. |
| Warning Device | Provide light, horn or voice warning if gear fails to extend and lock |
| Procedure | Provide instructions and training for: <br> 1) Manual hand crank operation to lower landing gear <br> 2) Landing gear-up type of landing |

This example shows the benefits gained by mitigations levels and demonstrates the relative weakness of the lower-level precedence options. Although it is possible to eliminate the hazard by replacing the retractable landing gear with fixed gear, this is not a desirable option for a high-speed aircraft. Obviously, the best solution is to improve the reliability of the landing gear through the use of design redundancy and high-reliability parts. A backup hand crank may be okay, but less desirable (and possibly less reliable) than a functional redundant design. The warning device and procedures for this example are not desirable because this would be an emergency situation, which the mitigations are trying to prevent.

12.6 System Safety Designed Layers of Defense

The overall goal of system safety is to develop a system or product that provides acceptable minimum potential mishap risk to the system users and bystanders. The basic system safety philosophy for achieving this goal is to design the system to provide three integrated levels of safety defense. These levels of defense are engineered into the system design by identifying three specific categories of hazards and mitigating their risk. This philosophy is summarized by the following three system safety precepts which revolve around three known categories of safety concern:

1) Design the system to operate safely during normal operating conditions
2) Design the system to safely operate and respond to abnormal operating circumstances
3) Design the system to provide survival protection from plausible mishaps

Although the basic system safety philosophy appears simple, actual implementation requires the work of a dedicated System Safety Program (SSP) that applies the System Safety process. Each of these safety philosophy precepts is discussed in further detail in order to provide an understanding of the considerations involved. This layered safety protection approach is analogous to an onion skin, as conceptualized in Figure 12.3. In this design safety approach one layer is built on top of another, providing three integrated levels of safety protection. If one layer fails then the next layer is intended to provide protection.

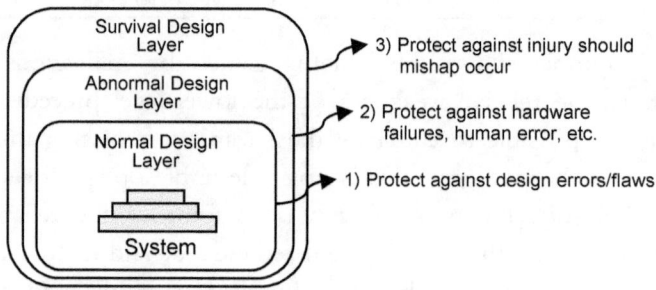

Figure 12.3 – Layered Design Safety Protection

The *Normal* design layer considers that the system works as designed under normally expected operating conditions (i.e., no hardware failures or abnormal conditions). The objective is to ensure the system is designed as intended, without the presence of possible design errors/flaws, software errors, sneak paths, integrations errors, etc. that could results in hazards.

The *Abnormal* design layer considers that the system responds safely to abnormal or unexpected operating conditions, such as hardware failures, human performance errors and/or abnormal environmental conditions. This is the most difficult layer to implement because system safety must anticipate and design around potential failures for all modes of operation. The objective is to ensure the system is designed to operate safely under fault conditions; anticipate the unexpected and counter through design safety features. It's a fact of life that hardware items will eventually fail; therefore it is imperative that the design take this into consideration for continued safe operation.

The *Survival* design layer considers user protection assuming the Normal and Abnormal levels of protection could fail, or extra user protection is desired (e.g., seat belts, airbags, crash helmets, etc.). Even after system hazards have been adequately mitigated it is often desirable to provide for user protection as an added margin of safety.

## 12.7 Guidelines for Design Safety Requirements

Each system safety program (SSP) should develop its own unique set of design safety requirements that apply specifically to the system involved. The purpose of a design safety requirement is to ultimately eliminate or mitigate a hazard. Safety requirements must provide value in the design guidance they provide; if they do not help the safety or the design organizations they do not provide a benefit. Safety requirements should give system developers and users' confidence that the particular safety concern will be safe when the requirement is followed.

When establishing DSFs it is recommended that consideration be given to the following design safety requirements, which are only a minimal list of example requirements and not a complete list. The design safety requirements in this section can be applied to many different types of systems. The requirements are divided into categories; however, they may be applicable across category lines.

12.7.1 Generic Safety Guidelines

- Establish *initial* safety design requirements from a review of applicable standards, specifications, regulations, design handbooks, lessons learned, and other sources of historical design guidance.
- Develop *derived* safety requirements needed to eliminate/mitigate hazards as the hazards are identified.
- Identify Safety-Critical Hardware Components (SCHCs) and ensure they are clearly identified on all associated specifications, drawing and documents.
- Identify Safety-Critical Computer Software Components (SCCSCs) and ensure they are clearly identified in the software requirements specifications and documents.
- Identify all components, functions, tasks and interfaces that can contribute to a hazard resulting in Catastrophic or Critical consequences as safety-critical.
- Ensure redundant designs are implemented, considering all factors that could result in their simultaneous failure, such as common cause failures.
- Physically separate and protect elements of redundant and backup systems to the greatest extent practical to minimize compromise by common cause failures.
- If the use of hazardous materials is necessary, select those that present the minimum risks, considering all operational phases and environmental conditions, including disposal. Document the types of hazardous materials used in the system and system equipment.
- Design and locate equipment so that required access during all operations, including maintenance on and off the system, minimizes personnel exposure to hazards.
- Design to minimize potential environmental exposure hazards by considering the impact of all possible environmental conditions, including adverse and extreme environments. Environments include natural and man-made environments (e.g., RF radiation).
- Design to eliminate hazards from foreign objects if possible. For those hazards that cannot be eliminated, provide safety measures to ensure the hazard risk from foreign objects is acceptable.
- Minimize the use of explosives and pyrotechnics as actuating devices.
- Design control system displays and malfunction detection systems to provide adequate warning of all safety-related malfunctions.
- Include suitable Warnings, Cautions, Advisories and special instructions in applicable technical data to address all procedurally controlled hazards.

- Ensure that system safety hazard mitigation methods are properly incorporated into the operator, user, and diagnostic manuals as specified by the corresponding system safety requirements.
- Design the system to operate safely in planned and expected modes of operation, as well as reasonable unexpected modes of operation.
- Design the system to operate safely within known and expected environments, as well as reasonable unexpected environments.
- Design the system to prevent unintentional and inadvertent commands, particularly those that are safety-critical and safety-related.
- Design the system to eliminate single-point failures (SPFs) for high (catastrophic) and serious (critical) risks. For those SPFs that cannot be removed, provide safety measures to ensure the hazard risk is acceptable.
- Design the system to eliminate common mode failures (CMFs) and common cause failures (CCFs) for high (catastrophic) and serious (critical) risks. For those CMFs and CCFs that cannot be removed, provide safety measures to ensure the hazard risk is acceptable.

## 12.7.2 Software Safety Guidelines

- Safety-critical system software should not include unintended/non-required functionality.
- The program shall ensure that COTS software is assessed for safety as a component in the system environment and not as an external element.
- COTS software is not exempt from system safety review and system hazard analysis.

## 12.7.3 Unmanned and Robotic Systems Safety Guidelines

- Design the system to perform only valid commands issued from a valid authorized operator, source or entity.
- Design the system to provide, obtain and/or utilize situational awareness information to support safe operations.
- Design the system to inform the state and/or mode of the system to the operator.
- Design the system to indicate if it is in a safe or unsafe state.
- Design the system to only allow states and modes changes that are safe.
- Design the system to prevent hazardous system mode combinations or transitions.
- Design the system to initialize/re-initialize in a known safe state.

- The system shall be designed to support options for operational or emergency contingencies.
- The system shall be designed to allow for safe and graceful degradation of the system upon critical failures within the system.
- The system shall be considered unsafe until a safe state can be verified.

## 12.7.4 Graphical User Interface (GUI) Safety Guidelines

The link between a system and a system operator (i.e., user) is the user interface (UI). The UI is the set of controls and displays via which a system user operates the system, be it a digital watch, a VCR, an automobile or a nuclear power plant. A graphical user interface (GUI) is a type of user interface that allows humans to interact with computers and electronic devices using graphical images on a computer screen rather than text based commands or mechanical devices. The GUI consists of graphical elements such as windows, menus, radio buttons, check boxes and icons, typically known as *widgets*. The GUI may employ a pointing device, mouse, touch screen and/or a keyboard. A poorly designed GUI can have can have a varying range of significance and consequences based on system complexity and safety critically. For example, designing a poor GUI for a stop watch may have trivial consequences, but a poorly designed GUI for a safety-critical function in a commercial aircraft control system can have significant safety ramifications. The following are some example safety guidelines for developing safe GUI designs:

- Design the GUI implementation to eliminate or reduce user confusion.
- Keep the number of GUIs to a minimum. Since they are screen widgets rather than mechanical devices, there is the temptation to provide a GUI for everything imaginable. Too many GUIs can cause user overload and confusion, which may impact safety.
- Avoid user mode confusion in GUI design. Always provide a visible indication of which system mode the current GUI represents.
- Keep GUI modes and states to a minimum. Heavy use of modes/states often reduces the usability of a user interface, as the user must expend effort to remember current mode states, and switch between mode states as necessary.
- Design the GUI implementation to minimize user workload. This will enhance safety by reducing user overload and stress.

- Conduct user test groups during design to test proposed GUI designs. The tests should assess if the GUI design confuses the operator, overloads the operator, allows the operator to commit errors, etc. This will help ensure the user can effectively use the GUIs as designed, and that they do not introduce any safety concerns.
- Consider a combination of mechanical devices and GUIs in an implementation. Some controls are better left mechanical, particularly safety controls. Consider the reliability of mechanical versus GUI widget for needed safety controls.
- When window screens and menus are employed, do not allow SC or SR windows or menus to be overlaid upon by other windows or menus. SC and SR windows and menus should always stay on top.
- Ensure the data latency in GUI information is appropriate, particularly for SC/SR applications.
- Perform a variety of tests on a GUI design to detect possible overlooked design hazards. Tests to consider include:
  - Screen swipe test (does it enter data?)
  - Random pushing of buttons
  - The entering of random and out of tolerance data

## 12.7.5 Human Error Safety Guidelines

Most mishaps attributed to human error are usually the result of a design error, which is discovered when the cause-effect relationship is carried back to the source causes. The following are some safety guidelines for developing safe designs that avoid potential human error problems:

- Ensure human engineering design criteria, principles, and practices are incorporated to the maximum extent practical in the system design to ensure safe system operation, training, and maintenance.
- Design to minimize human error potential in the operation and maintenance of the system.
- Design to expect and protect against potential human error in safety-critical functions.
- Consider all human error types and their potential for safety problems, such as:
  - Slips – Consider the safety implications of a user erroneously performing a task. For example, when an operator erroneously presses the wrong button because several buttons are close together. Use guarded buttons and confirmation of action when necessary.
  - Omissions – Consider the safety implications of a user omitting a particular step (intentionally or unintentionally).

- ▪ Commissions – Consider the safety implications of a user performing a task out of sequence or an unrelated task (intentionally or unintentionally).
- Legibility – users must be able to easily read displays, etc.
- Comprehension – users must be able to easily understand information presented to them.
- Pictorial realism – displays must provide standard realism (e.g., thermometer – vertical; low to high temp at top).
- Consistency – use format of old familiar displays (avoid transition errors).
- Moving indicators should be compatible with the user's mental model (e.g., altimeter should move upward with increasing altitude).
- Use information redundancy when feasible, for example, a traffic light uses both color and position to provide information. Use both visual and auditory cues.
- Ensure similarity in different displays causes confusion.
- Do not determine the absolute level of a variable on the basis of a single sensory input (e.g. color, size, loudness).
- Make data and information easily accessible by the user; don't force the user to memorize information, especially safety related information.

# 13   COMMON MISTAKES IN SYSTEM SAFETY

13.1 Common Mistakes

Quite often a system safety program (SSP) can be improved after giving consideration to lessons learned from past mistakes on other programs and systems. The following are some useful lessons learned and mistakes to avoid when conducting an SSP:

- Not making system safety a value-added process that is fully implemented and supported. Hazards and risk can be eliminated or mitigated only when a recognized and supported SSP is conducted.
- Assuming system safety is too expensive to perform. Safety is typically much more expensive when hazards have not been identified and eliminated/mitigated, thereby preventing mishaps.
- Not using a qualified and experienced system safety staff. Using an inexperienced SSP staff is almost like not even performing an SSP.
- Not having official company-approved and documented system safety policies and methods.
- Not making safety a true core value of the company and/or the project.
- Failing to involve the reliability and human factors disciplines in the system safety process.
- Failing to train system designers and technical area experts on the project in the system safety process, tools and methods. When the designers understand the system safety process, they can make better design decisions for safety.
- Allowing the SSP to become an isolated organization, rather than working as partners for success with the other program organizations and disciplines.

- Making system safety a check-the-box effort, rather than a core value of the company. Not being fully dedicated to the complete system safety process.
- Not thinking in Failure Space as opposed to Success Space. System safety practitioners have to keep asking the question, "How can this fail and what happens when it fails?" Designers typically only look at system success and do not appreciate the failure side of a system.
- Failing to completely design-out, or eliminate, hazards whenever possible, even when the risk is already small.
- Thinking that a hazard is eliminated when the risk presented by the hazard is mitigated. The Hazard Triangle concept must always be kept in mind.
- Not performing a thorough, complete and correct HA.
- Not using the right HA method or tool.
- Not using experienced and knowledgeable staff.
- Not providing adequate resources for the HA.
- Using a generic HA rather than a system specific one.
- Not probing deep enough into all of the systems hazard sources.
- Not writing hazards in their full context, including HS, IM, and TTO.
- Writing hazards and assessing risk at the wrong level in the system hierarchy.
- Not thoroughly understanding hazard theory and HA methods.
- Not providing training to the system safety staff in order to improve their knowledge and skills.
- Failing to perform a system oriented safety analysis on COTS items.
- Complacency in performing the system safety process.
- Failing to listen to feedback.
- Failing to speak out and pursue safety concerns in the face of opposition.
- Not allowing the system safety manager to have a direct voice with the program manager to resolve safety concerns in the face of opposition.
- Failing to establish a lessons-learned database.
- Failing to research and apply lessons learned.
- Failing to implement and utilize an incident field monitoring system.
- Allowing business monetary decisions to dominate over the ethics of developing safe products.

We are doomed to repeat past mistakes if we do not keep a record of them and continuously study them and apply lessons learned in current system safety programs.

# 14  SOFTWARE SAFETY

14.1 Software Safety Introduction

Software safety (SwS) is the process of developing safe software, software that executes within a system context and environment with an acceptable level of potential mishap risk. This means the software will not cause or contribute to any system mishaps or prevent system design safety mechanisms from performing correctly within acceptable bounds. The SwS process is the intentional and planned application of management and engineering principles, criteria, and techniques to develop software safe for use in a specific system. It should be noted that SwS is not the same as software reliability or software quality assurance, and it cannot be achieved solely through these processes. The SwS methodology involves an independent standalone process that must be integrated with the system safety and software development processes.

The scope and coverage of SwS includes computer software, firmware and programmable logic arrays. SwS is primarily concerned with application software developed as part of a system development program. However, due to the permeating nature of software, SwS must also consider operating systems, compilers, software tools and reused software, including any form of COTS software that is utilized in the system.

The use of software in a system presents a paradoxical situation. On the one hand, software provides many benefits to the user, including increased flexibility and speed, greater accuracy, and enhanced system control. On the other hand, that same software can create unforeseen hazards that are not always well understood or easily recognizable. Software definitely increases the potential mishap risk of a system and requires significantly more effort to ensure safety. Technical advancements have made the digital computer both less expensive and more powerful in capability. The resultant effect is that computers now dominate the control of system functions, and the software that operates these computers has become a major system element that presents potential mishap risk.

Software is becoming so pervasive that it not only impacts the safety of military weapon systems with significant adverse consequences, but it also impacts the safety of everyday life with its incorporation into products and systems such as microwaves, cell phones, traffic lights, banking, home security, air travel, rail travel, automobiles, etc. Software can, and already has, caused mishaps with automobiles, medical equipment, spacecraft, trains, aircraft and weapon systems. Therefore, SwS is an important factor in system design and development which cannot be ignored.

The following are some SwS principles that have been established from the unique characteristics of software. These principles are useful when applying the SwS process and when identifying software-related hazards:

- Software by itself is not hazardous; it is only hazardous in a system when performing system functions involving hardware.
- Hardware causes the damage in a mishap (i.e., explosives, radiation equipment, flight controls, chemicals, etc); however, software can be an initiating or controlling factor.
- When identifying software-related hazards, look for safety-related hardware-software relationships, since software can contribute to hazards only via hardware.
- SwS requires multiple perspectives: system, hardware interfaces, human interfaces and functions.
- Software code has no concrete failure modes as does hardware; software does have functional failure modes.
- Hardware faults can induce software functional failures (modifies or fails software intent).
- Software can have errors and still function (safely or unsafely).
- Not all software errors are safety-related.
- Software hazard risk is difficult to quantify (can't quantify software errors or functional failures).
- Software always works exactly as coded, but complexity makes comprehension difficult.
- Sometimes software does more than intended or expected (this is a safety concern).

14.2 Software Safety Process

In theory, the Software Safety (SwS) process is a subset of the system safety process and similar in methodology. However, due to the unique characteristics and nature of software, the SwS process deviates slightly and takes a more diverse approach. Whereas system safety is risk-based, SwS is assurance-based (also sometimes referred to as integrity-based). Hardware safety focuses primarily on mitigating hazard risk to an acceptable level. In the case of software, actual hazard risk cannot be calculated, thus acceptable risk is nebulous and is based on a diverse SwS assurance process. The SwS assurance process focuses on functional safety (design) assurance and software development assurance.

Although SwS generally applies the system safety process that was established for hardware safety, software must be treated slightly differently due to the complexity, special attributes and unique nature of software. For example, when software is involved in a hazard it is not possible to obtain the failure rate of potential software causal factors (as is done with hardware) and therefore a risk likelihood value cannot be computed for a risk assessment. In addition, when a software-related hazard is identified, it is often difficult or impossible to determine if specific causal factors actually exist in the complex and abstract code modules. Software hazard analysis is not sufficient for SwS; extensive testing must also be performed to determine if identified hazards can occur, or if previously unidentified hazards exist. The development of safe software involves a mix of hazard analysis, design safety requirements scrutiny, safety-critical function (SCF) scrutiny, significant testing and the use of rigorous software development methods and tools.

The unique nature of software causes the existence of two enigmas associated with SwS, which in turn cause the need for a more diverse approach in order to ensure adequate safety of software. The two stumbling blocks in SwS are:

1) Software functional failures (and hazards) can be postulated, but they cannot usually be definitively proven by specific identifiable causal factors in the software.
2) Failure rates cannot be determined for software functional failures, therefore hazard risk cannot be calculated for software-related risk assessments.

Because of these two software safety dilemmas, it's evident that software presents potential mishap risk that is unknown and cannot be precisely determined. Therefore, the most pragmatic way to ensure that software is safe is by applying a bilateral safety approach consisting of: (1) software functional coverage and (2) software developmental coverage. The software functional coverage scheme focuses on the functional design to provide hazard identification and mitigation assurance and SCF identification and assurance. The software development coverage scheme utilizes the software development process to assist in the forced focus on specific development tasks that ensure higher-quality software that is presumably safer. The theory is that if the software is developed to a specified set of rigorous requirements, analyses, tests and development procedures the resulting product will present acceptable safety risk. When all the appropriate hazard mitigation tasks and software development tasks are successfully performed, the overall software mishap risk is "judged" to be acceptable. This bilateral approach is a strategy intended to broadly cover all aspects of software that can impact safety. This software safety scheme provides a "presumed level of assurance" that the software has received complete safety coverage and the risk presented by the software is deemed acceptable. SwS assurance requires visibility of both the product and the process. It should be noted that these SwS tasks do not provide a quantitative estimate of the potential mishap risk associated with the software. What this approach does provide is a level of confidence that the software can be considered as being safe. Software-related hazards can be accepted for risk based on the conclusions drawn from the safety case, where the safety case is built upon the results of both the functional and developmental completion evidence.

All of the elements in this approach are interrelated and dependent upon each other. These steps do not imply any sort of order or sequence; the necessary steps should be performed as it makes sense for the project. The basic steps include:

1) Perform system hazard analysis to identify hazards. Identify both hardware and software causal factors. The software causal factors lead to design mitigation requirements in the form of system safety requirements (SSRs). This step also leads to the identification of software criticality levels (SCLs), which will impact the Level of Rigor (LOR) tasks performed by the software development effort.

2) Perform functional hazard analysis to identify safety-critical functions (SCFs). This also leads to the identification of safety-critical (SC) code modules and design SSRs. SC requirements are tagged in the requirements tracking system for close scrutiny and testing. This branch also supports establishing SCLs for the software modules.

3) Identify the appropriate design safety requirements and apply them to the software. These requirements will include baseline, derived and generic requirements. Baseline safety requirements stem from the contractual functional design requirements. Derived safety requirements are SSRs established to mitigate hazards and potential safety issues. Generic software safety requirements are industry known guidelines and requirements. The generics are very basic and general software requirements that have been found to be useful in helping to assure safe software design. The software design group implements the generics and completes a checklist that provides evidence of completion.

4) Establish the LOR tasks required for each SCL. Perform the LOR tasks on the software modules and document the evidence of successful completion. This effort is done primarily by the software development organization.

To accept the risk provided by the software, a safety case must be developed which provides assurance that the software is considered acceptably safe. The safety case for SwS is largely based on the following evidence, as a minimum:

- Safety-related hazards have appropriate mitigation methods, which are in the form of SSRs.
- Hazard mitigation SSRs have been successfully tested.
- SCFs have been documented and the appropriate SSRs have been established to protect these functions against adverse behavior.
- Software design requirements have been reviewed and those that are SC have been tagged as SC.
- SCF SSRs and SC SSRs have been successfully tested.
- All LOR tasks have been successfully performed.
- All levels of requirements can be traced to their roots (and hazards) and each is fully tested.

The SwS assurance process focuses on the combination of: a) functional safety (design) assurance and b) software development assurance as depicted in Figure 14.1. The functional safety aspect focuses on the design and software related hazards. The software development assurance aspect focuses on applying a rigorous software development process. Those software modules that are most safety-critical receive a higher level of design and test rigor.

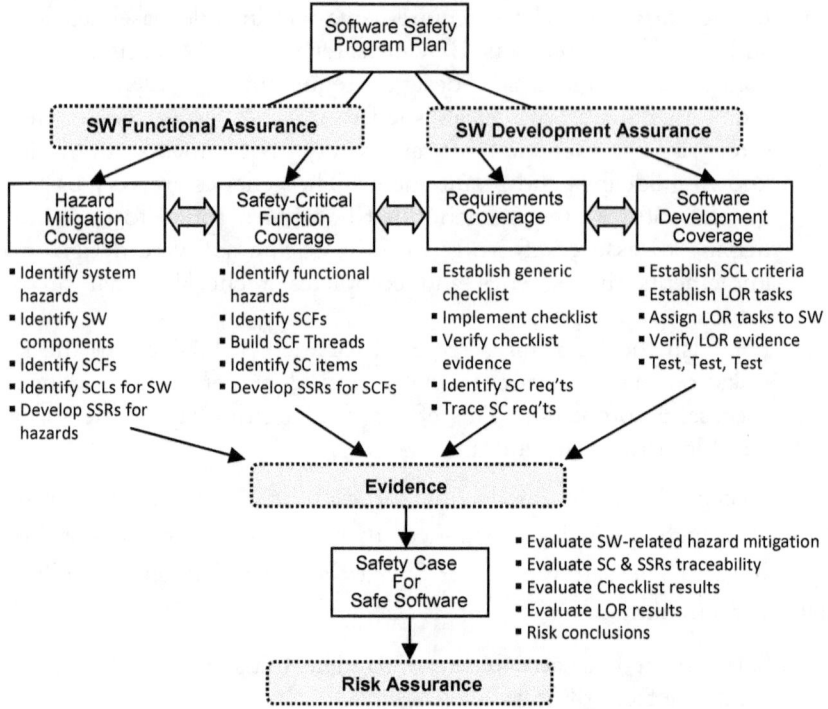

Figure 14.1 – Software Safety Process

SwS involves performing a system safety process on software to ensure the software will perform its function with an acceptable level of mishap risk. SwS is not just a software issue, it is a systems issue. Software related hazards must be identified, understood and mitigated to an acceptable risk level, considering that software interfaces with hardware, humans and other software. SwS is a specialized subset of system safety, and is handled slightly differently than hardware due to the unique nature and characteristics of software. Optimally, Software safety is an integral aspect of the overall SSP and the methodology is documented in the SSPP. This approach provides an integrated and effective method for the identification and control of software contributions to system level hazards, and it also minimizes any impact to the overall program cost and schedule. Detailed SwS analyses and test verification activities provide evidence that safety risks associated with the use of safety-critical software are mitigated or low risk. Software's contribution to system level hazards, and hazard mitigation, must be assessed within a structured and disciplined SSP.

# 15  SYSTEM SAFETY VALIDATION

## 15.1 Validation

Individuals unfamiliar with the system safety processes often question the merits of system safety, with questions such as:

- Are the claims made regarding system safety benefits realistic?
- Where is the data substantiating these claims?
- Why isn't a standard design process equal or sufficient?
- How can you immediately see what you are buying (in terms of safety)?

These are legitimate questions which should be answered in order to validate system safety concepts and processes. Cogent evidence of system safety success is often difficult to obtain in a timely manner. Achieved system safety is a somewhat invisible and intangible quality or attribute. We know a system is safe when mishaps are not occurring and it is unsafe when mishaps are occurring, which is demonstrated by operation over time. We also know that by eliminating hazards the system is safer, therefore a formal system safety effort must be spent to identify and eliminate hazards.

Tracking accident/incident rates over time is a lagging safety indicator that is inconvenient for proactively measuring safety gains. Also, an absence of mishaps does not necessarily equate to a safe system. It would be much more efficient and effective to know immediately that a specific design architecture prevented specific potential mishaps from occurring. This is typically achieved by identifying hazards and then eliminating them when possible. Those hazards that cannot be eliminated are controlled by design safety features, and the amount of control is measured by the metric of risk. Risk is a useful safety yardstick; even though it is an incorporeal quality it provides a valid measure of safety and safety improvement.

The best way to confirm the system safety process is to look at the safety results of systems from several different viewpoints, such as the following:

- Successful system safety programs
- Successful hazard analysis programs
- Lessons learned from preventable mishaps

An important aspect of system safety is confirming that a system safety program (SSP) did prevent mishaps and save lives, without waiting until the end of the system's lifecycle (when everyone has forgotten). There are many examples of successful system safety programs, but many have been forgotten or not recognized, such as the examples discussed below. Another way to demonstrate the payoff of system safety is to evaluate past mishaps and determine if they could have been prevented by an SSP, as discussed below.

## 15.2 Successful System Safety Programs

Documenting a successful SSP is difficult because the results are not known until the system has been in operation for some time without mishaps. The following two systems are examples where an extensive SSP was applied and the efforts have been shown over time to have been extremely successful and worthwhile.

### Minuteman Missile System

The Minuteman is a U.S. nuclear intercontinental ballistic missile. The Minuteman Missile System is a large and complex silo-based network of intercontinental ballistic Minuteman missiles with nuclear warheads. An extensive SSP was applied by the Boeing Company during the development of the Minuteman Missile System, initially deployed in 1962. Safety was a major design factor, and system safety was utilized to design-in safety right from the start of the program. As a result, no major mishaps have occurred during the life of the system, which is still in operation. The system safety process was essentially created, proven and refined on the Minuteman system.

### Morgantown PRT

The Morgantown Personal Rapid Transit (PRT) is an automated people mover system located in Morgantown, West Virginia. The system connects the three Morgantown campuses at the University of West Virginia, as well as the downtown area. It was developed by the Boeing Company and entered service in 1975. An extensive system safety program (including software safety) was applied during development of the system. As a result of the system safety effort, not a single incident of serious injury and/or death has occurred on the Morgantown PRT system, which is still in operation.

## 15.3 Successful Hazard Analysis Programs

The following is an example where an extensive SSP was successfully applied. A detailed safety analysis showed where a design change (during the design phase) was necessary in order to eliminate an identified hazard.

## B-1A TFR

The B-1A bomber has a Terrain Following Radar (TFR) system that can automatically fly the aircraft at a very low and constant altitude above the terrain. In addition, the TFR has a built-in safety system that can sense an impending obstacle on a collision course with the aircraft, and it generates an automatic fly-up command to avoid the obstacle. The fly-up system is triple-redundant, such that failure of all three branches is necessary before the safety system fails. The Boeing Company developed the B-1A avionics, which utilized a TFR system that was provided as a commercial off-the-shelf (COTS) item by the Air Force. An extensive SSP was performed, and hazard analysis showed that the fly-up system was safety-critical. A fault tree analysis (FTA) was performed on the fly-up system to ensure it was fully triple-redundant. The FTA revealed that the system was not fully redundant as presumed, that all three independent branches were fed by one common power supply. If any component in the power supply failed, it acted as a common cause failure that caused failure of all three redundant branches. A design change was made to correct this condition, and no aircraft have been lost due to this potential hazard. The analysis was documented in Boeing document D229-10316-1, B-1A Terrain Following Radar Safety Study, 1975.

## 15.4 Lessons Learned From Preventable Mishaps

The results of system safety are often not visible when the SSP has been successful in preventing mishaps; unfortunately, a prevented mishap is not a visible or quantifiable metric. This phenomenon tends to under-rate and under-value the benefits of the SSP. However, when mishaps do occur, system safety (or the lack thereof) becomes very visible. Evaluation of the following mishaps helps to underscore how these systems could have benefitted from a thorough and systematic SSP.

DC-10 Mishap

United Airlines Flight 232 was a scheduled flight between Denver and Philadelphia via Chicago. On July 19, 1989, the Douglas DC-10 (Registration N1819U) suffered an uncontained failure of its number 2 engine (mounted in the tail), which destroyed all three of the aircraft's hydraulic systems. With no controls working except the power levers for the two remaining engines, it broke up during an emergency landing on the runway at Sioux City, Iowa, killing 110 of its 285 passengers and one of the 11 crew members. While the plane was in a shallow right turn at 37,000 feet, the N1 stage fan disk of its tail-mounted General Electric CF6-6 engine broke in two. The fan cowling was blown off and pieces of the engine penetrated the aircraft tail section in numerous places, including both horizontal stabilizers. Pieces of shrapnel went through the right horizontal stabilizer, severing the lines of all three hydraulic systems, allowing the fluid to drain away. Since the failure of all three hydraulic systems was considered extremely unlikely by the aircraft engineers, there was no backup means of safely controlling the aircraft.

> Lesson: The three redundant hydraulic lines were likely considered a design safety feature to prevent loss of hydraulic power to the flight control surfaces. In retrospect, had a more thorough system safety analysis been performed, it might have been recognized that the three hydraulic lines were placed in the same location and were thus susceptible to a single common cause failure that could wipe out the intended safety feature. The triple-redundancy design may be considered a minimum design standard that was presumed to provide adequate safety. This example demonstrates that more safety effort is required than just meeting minimum standards in order to ensure that specific hazards are identified and addressed (such as common cause failure analysis in this case).

Pinto Mishap

The Ford Pinto is a subcompact automobile that was produced by the Ford Motor Company for the 1971–1980 model years. Controversy followed the Pinto after 1977 allegations that the Pinto's structural design allowed its fuel tank filler neck to break off and the fuel tank to be punctured in a rear-end collision, resulting in deadly fires. Critics alleged that the vehicle's lack of reinforcing structure between the rear panel and the tank meant the tank would be pushed forward and punctured by the protruding bolts of the differential. Ford was allegedly aware of the design flaw, refused to pay for a redesign, and decided it would be cheaper to pay off possible lawsuits for resulting deaths. A 1972 accident that killed one person and severely burned another went to court, and the plaintiffs were awarded compensatory damages of $2.5 million and punitive damages of $3.5 million against Ford, partially because Ford had been aware of the design defects before production but had decided against changing the design. In 1978, Ford initiated a recall, providing a plastic protective shield to be dealer-installed between the fuel tank and the differential bolts, another to deflect contact with the right-rear shock absorber, and a new fuel-tank filler neck that extended deeper into the tank and was more resistant to breaking off in a rear-end collision.

> Lesson: Had a formal SSP been implemented, applying the ethics of system safety, the design hazard would have been eliminated during product development stage (it apparently had been recognized but ignored due to a cost vs. liability tradeoff). In addition, had an effective safety field monitoring system been in place that quickly resolved identified safety issues, the recall could have occurred much sooner and without litigation.

Therac-25 Mishap

The Therac-25 is a classic example of poor engineering combined with the lack of a system safety program. The Therac-25 is a radiation therapy machine produced by Atomic Energy of Canada Limited (AECL). It followed after the Therac-6 and Therac-20 units. It was involved in at least six accidents between 1985 and 1987, in which patients were given massive overdoses of radiation, approximately 100 times the intended dose. The accidents occurred when the high-power electron beam was activated instead of the intended low power beam, and without the beam spreader plate rotated into place. The machine's software did not detect that this had occurred, and therefore did not prevent the patient from receiving a potentially lethal dose of radiation. A commission concluded that the primary reason for the accidents should be attributed to bad software design and development practices, and not explicitly to several coding errors that were found. In particular, the software was designed so that it was realistically impossible to test it in a clean automated way. The software was written in assembly language, which typically requires more attention for testing and good design, however the choice of language by itself is not listed as a primary cause in the report. The system also used its own operating system. Researchers who investigated the accidents found several contributing causes to the accidents. For example, the AECL did not consider the design of the software during its assessment of how the machine might produce the desired results and what failure modes existed (i.e., no system safety program). The AECL did not have the software code independently reviewed. When the system noticed that something was wrong it halted the X-ray beam, but then it merely displayed the word "MALFUNCTION" followed by a number from 1 to 64. The user manual did not explain or even address the error codes, so the operator pressed a key to override the warning and proceed anyway. The AECL personnel, as well as machine operators, initially did not believe complaints; possibly due to overconfidence. The AECL had never fully tested the Therac-25 with the combination of software and hardware until it was assembled at the hospital. The failure only occurred when a particular nonstandard sequence of keystrokes was entered. This sequence of keystrokes was improbable, so the problem did not occur very often and went unnoticed for a long period of time. The design did not have any hardware interlocks to prevent the electron-beam from operating in its high-energy mode without the target plate in place, whereas previous models did have hardware interlocks.

> Lesson: Had an SSP, along with a software safety program, been performed, the appropriate hazards would have been identified and eliminated during the design phase. Deaths, injuries and litigations could have easily been prevented. This is a safety-critical device where no apparent concern was given to design-for-safety or the identification of hazards. The Therac-25 accidents highlight the dangers of software control of safety-critical systems, and this mishap has become a standard case study in software safety engineering.

In conclusion, system safety has been validated as a successful process for identifying and eliminating hazards, as shown by the examples provided above. In addition, it should be mentioned that system safety requires, in part, proactive field monitoring of systems to identify when systems are misbehaving, in order that system safety can be applied to correct the problem.

# APPENDIX A – EXAMPLE HAZARD ANALYSIS

A.1 Gas Furnace System

To help demonstrate the hazard analysis (HA) process a HA is performed on an example hot water heater system. Figure A.1 shows two views of a generic gas house hot water heater: view A is a pictorial view of the heater and view B is a simplified system diagram of the heater system. Note that the simplified diagram serves as an invaluable aid in recognizing hazards.

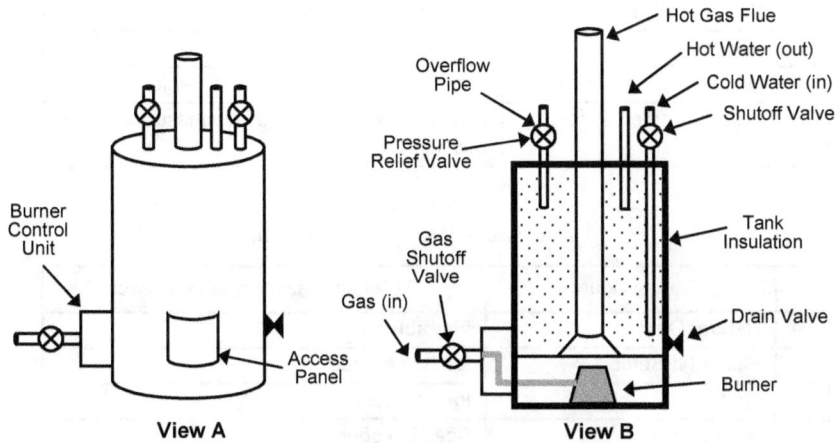

Figure A.1 – Gas Water Heater Diagrams

The first step in the HA is to develop a system equipment list or a system hierarchy table for the system. The system safety analyst must ensure that all of the system equipment is covered by the HA. Table A.1 contains an equipment list for this system.

The second step is to identify the energy sources in the system that could act as a Hazard Source. Table A.2 contains a list of the energy sources for this system.

The next step is to identify the safety-critical functions in the system that could act as a Hazard Source. Table A.3 contains a list of the major functions for this system.

If the information provided in Tables A.1 through A.3 is not supplied by the program, it is then incumbent upon the system safety analyst to develop this information.

## Table A.1 – System Equipment List

| | System Item | Purpose |
|---|---|---|
| 1 | Tank | Provides containment for hot water |
| 2 | Gas Burner Unit | Provides heat to produce hot water |
| 3 | Burner Control Unit | Controls gas burner |
| 4 | Gas Shutoff Valve | Shuts off gas to unit in emergency or maintenance |
| 5 | Gas Pipe | Provides and contains gas to unit |
| 6 | Tank Drain Valve | Drains water from tank |
| 7 | Pressure Relief Valve | Relieves over-pressure from tank |
| 8 | Hot Gas Flue | Provides exit for burned gas and air |
| 9 | Water-In Pipe | Provides cold water into tank |
| 10 | Water-Out Pipe | Provides hot water out of tank |
| 11 | Water-In Shutoff Valve | Shuts off colder water input to tank |
| 12 | Water Overflow Pipe | Provides exit for water after pressure release occurs |

## Table A.2 – Energy Source List

| | Energy Source | Potential Hazard Source Impact |
|---|---|---|
| 1 | Natural Gas | Fire source |
| 2 | Flame (in Burner) | Ignition source |
| 3 | Igniter | Ignition source |
| 4 | Hot Water | Scalding source |
| 5 | Hot Flue Gas | Skin burn source |
| 6 | Water Pressure | Tank rupture source |

## Table A.3 – System Function List

| | Functions | Purpose |
|---|---|---|
| 1 | Control gas flow to Burner | Provide gas to burner when required |
| 2 | Ignite gas | Ignite the gas |
| 3 | Burn gas | Continue to burn gas |
| 4 | Cold water in | Provide for cold water input |
| 5 | Hot water out | Provide for hot water outflow from tank |
| 6 | Shutoff gas input | Stop gas input to burner |
| 7 | Shutoff cold water input | Stop cold water input |
| 8 | Pressure release | Relieve over pressure situation |
| 9 | Vent burned gas-air | Remove burned gas-air from burner |
| 10 | Drain tank | Remove water for cleaning |

The PHL analysis is essentially a brain storming session. The intent is to identify hazard sources and generic hazards. The information from the PHL will be used to generate individual hazards in the PHA. Tables A.4 through A.6 are PHLs on the system hardware, energy sources and functions, respectively. Note that the acronym D/I stands for death and/or injury. TLM stands for Top Level Mishap which is merely a generic mishap category used to categorize and organize hazards.

## Table A.4 – PHL of System Hardware

| Preliminary Hazard List Analysis | | | | |
|---|---|---|---|---|
| System Element Type: *System Hardware* | | | | Page 1/3 |
| No. | System Item | Hazard Source | Generic Hazard | TLM |
| 1 | Tank | Water pressure<br>Water (liquid) | Tank ruptures from overpressure<br>Water leak damages items in area | Damage<br>Damage |
| 2 | Burner Unit | Flame<br>Igniter sparks<br>High temperature | Ignites other gases in area<br>Ignites other gases in area<br>Injures user during maintenance | Fire<br>Fire<br>User D/I |
| 3 | Burner Control Unit | Natural gas | Fails to provide gas to Burner<br>Fails to stop gas flow to Burner | Damage<br>Fire |
| 4 | Gas Shutoff Valve | Natural gas | Unable to close shutoff valve | Damage |
| 5 | Gas Pipe | Natural gas | Gas leak; fire; explosion | User D/I |
| 6 | Tank Drain Valve | Water | Water leak causing water damage | Damage |
| 7 | Tank Pressure Relief Valve | Water pressure | Valve fails to release pressure; rupture | Damage |
| 8 | Hot Gas Flue | Hot gas (CO)<br>Carbon Monoxide | Hot pipe ignites materials<br>CO poisoning | Fire<br>User D/I |
| 9 | Water-In Pipe | Water | No water input; no hot water<br>Water leak | None<br>Damage |
| 10 | Water-Out Pipe | Hot water | Water leak | Damage |
| 11 | Water-In Shutoff Valve | Water | Unable to close shutoff valve | Damage |
| 12 | Water Overflow Pipe | Hot water | Pipe is clogged, fails to release high pressure | Damage |

## Table A.5 – PHL of System Energy Sources

| | Preliminary Hazard List Analysis | | | |
|---|---|---|---|---|
| System Element Type: *System Energy Sources* | | | | Page 2/3 |
| No. | System Item | Hazard Source | Generic Hazard | TLM |
| 1 | Natural Gas | Fire<br>Explosion<br>Carbon Monoxide | Gas leak ignition results in fire<br>Gas leak ignition results in explosion<br>Incorrect burning produces CO | Fire<br>Fire<br>User D/I |
| 2 | Flame (in Burner) | Flame | Ignites other gases in area | Fire |
| 3 | Hot Water | Water Temperature | Hot water injures user | User D/I |
| 4 | Igniter | Sparks | Ignites other gases in area<br>Fails to ignite flowing gas (no shutoff) | Fire<br>User D/I |
| 5 | Hot Flue Gas | High temperature gas | Ignites materials | Fire |
| 6 | Water Pressure | High pressure in tank | Pressure not released; tank ruptures | Damage |

## Table A.6 – PHL of System Functions

| | Preliminary Hazard List Analysis | | | |
|---|---|---|---|---|
| System Element Type: *System Functions* | | | | |
| No. | System Item | Hazard Source | Generic Hazard | TLM |
| 1 | Control gas flow | Gas control fails | Fire | Fire |
| 2 | Ignite gas | Fails to ignite gas | Gas leak; fire | Fire |
| 3 | Burn gas | Fails to burn gas | Gas leak; fire; CO poisoning | Fire |
| 4 | Cold water in | No water in; Leaks | Water leak; no hot water | Damage |
| 5 | Hot water out | No water out; Leaks | Water leak; no hot water; tank overpressure | Damage |
| 6 | Shutoff gas input | Fails open | Gas leak; fire | Fire |
| 7 | Shutoff water input | Fails open | Water leak | Damage |
| 8 | Pressure release | Fails closed | Tank overpressure | Damage |
| 9 | Vent burned gas | Leaks | CO poisoning | D/I |
| 10 | Drain tank | Fails open | Water leak | Damage |

The Preliminary Hazard Analysis (PHA) worksheet can take many different forms and formats. The PHA form in this example has been abbreviated utilizing fewer columns in order to fit the book format. Tables A.7 and A.8 are example PHAs performed on the system design.

Table A.7 – PHA (Page 1 of 2)

| No. | Hazard | Causes | Effects | Risk |
|-----|--------|--------|---------|------|
| Preliminary Hazard Analysis | | | | |
| 1 | Tank<br>Tank ruptures from overpressure | Pressure buildup and relief valve fails | Tank rupture; injury; system loss | H |
| 2 | Water leak damages items in area | Tank material failure | Area water damage | L |
| 3 | Burner Unit<br>Ignites gases in area (gas leak; gasoline vapors; paint vapors) | Flame or igniter provide ignition source | Fire/ explosion; D/I; sys loss | H |
| 4 | Fails to burn gas | Burner failure | Gas leak; fire | H |
| 5 | Hot surface burns user during maintenance | Hot surface | User injury | L |
| 6 | Burner Control Unit<br>Fails to provide gas to Burner | Control unit failure | No heat | L |
| 7 | Fails to stop gas flow to Burner when flame is out; cannot ignite | Control unit failure | Gas leak; fire | H |
| 8 | Gas Shutoff Valve<br>Unable to close shutoff valve during an emergency, allowing gas to burner. | Valve fails closed | Potential fire/explosion | H |
| 9 | Gas Pipe<br>Leak in gas pipe allows gas in area. | Pipe failure | Potential fire/explosion | H |
| 10 | Tank Drain Valve<br>Valve fails open causing water leak that results in damage. | Valve fails open | Area water damage | L |
| 11 | Tank Pressure Relief Valve<br>Valve fails closed failing to release pressure when required, resulting in tank rupture. | Valve fails closed and overpressure occurs | Tank rupture; damage | M |
| 12 | Hot Gas Flue<br>Hot pipe ignites materials. | Materials touch pipe | Fire | H |
| 13 | Leaking flue allows CO into area. | Pipe failure or damage | CO poisoning resulting in D/I | H |

## Table A.8 – PHA (Page 2 of 2)

| No. | Hazard | Causes | Effects | Risk |
|---|---|---|---|---|
| | Preliminary Hazard Analysis | | | |
| 14 | Water-In Pipe<br>Pipe is plugged preventing entry of cold water. | Debris in water line | No water input; possible tank over heating | M |
| 15 | Pipe leaks allowing water drainage. | Pipe failure | Water damage in area | L |
| 16 | Water-Out Pipe<br>Pipe is plugged preventing exit of hot water. | Debris in water line or tank | No water output; possible tank overpressure | M |
| 17 | Pipe leaks allowing water drainage. | Pipe failure | Water damage in area | L |
| 18 | Water-In Shutoff Valve<br>Valve fails open and leaks water in area. | Valve fails open | Water damage in area | L |
| 19 | Water Overflow Pipe<br>Pipe is plugged preventing entry of cold water. | Debris in water tank | No water output; possible tank overpressure | M |
| 20 | Pipe leaks allowing water drainage. | Pipe failure | Water damage in area | L |

Note that on the PHA form the required HS-IM-TTO information is provided in the following columns:

- Hazard – identifies the HS and the basic hazard scenario
- Causes – identifies the hazard IMs
- Effects – identifies the hazard TTO

# APPENDIX B – EXAMPLE RISK RATING METHODOLOGY

B.1 Risk Rating Using the HRI Matrix

For rating and accepting hazard risk, the Hazard Risk Index (HRI) methodology is a proven and effective technique. This method provides for a good characterization of risk, which can be estimated qualitatively or quantitatively. It also provides a relatively simple methodology that is cost-effective to perform. This is a commonly used approach by system safety engineers in both defense and non-defense applications. It is recommended by MIL-STD-882D, Standard Practice for system safety and by ANSI/GEIA-STD-0010-2009, Standard Best Practices for system safety Program Development and Execution.

The HRI matrix has several intertwined purposes that are useful to system safety. The HRI matrix allows the system safety analyst to:

1) Perform a risk assessment that is simple, efficient and cost effective
2) Determine the potential mishap risk presented by a hazard
3) Communicate hazard risk from a common framework
4) Rank hazards by risk index level
5) Prioritize highest risk hazards requiring immediate mitigation attention
6) Know how much mitigation is necessary to lower the risk to an acceptable level
7) Identify the level of authority that can accept the residual risk presented by a hazard
8) Compare hazards by risk level across a program and across various platforms

The HRI Matrix also goes by other names; however, regardless of the name used, the matrices are essentially the same entity because hazard risk is the same entity as mishap risk. Some of the alternate names include: Risk Hazard Index (RHI) matrix, Mishap Risk Index (MRI) matrix and Mishap Risk Assessment Matrix (MRAM). Variations of the HRI Matrix are used in different industries and agencies.

B.2 HRI Concept

The HRI matrix establishes the relative level of potential mishap risk presented by an individual hazard. By comparing the calculated qualitative severity and likelihood values for a hazard against the pre-defined criteria in the HRI matrix, a level of risk is determined by a derived index number. The HRI matrix concept essentially involves one matrix and three tables, as depicted in Figure B.1.

The HRI matrix is the main component, based upon the combination of the hazard/mishap likelihood on one axis and hazard/mishap severity on the other axis. The hazard/mishap likelihood category is determined from the criteria stated in the Likelihood Table, and the hazard/mishap severity category is determined from the criteria stated in the Severity Table. The Risk Level table ranks each hazard into one of four risk levels (High, Serious, Medium or Low) based on the particular HRI matrix indices designated for the particular level.

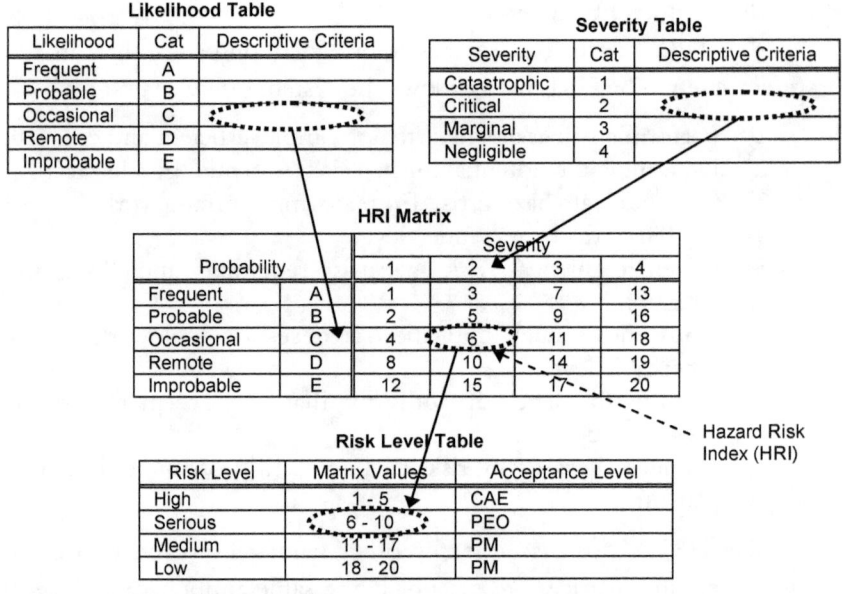

Figure B.1 – HRI Concept

The relative risk index (or HRI) is derived from the matrix cell resulting from the intersection of the likelihood and severity axes that represent a particular hazard. The HRI matrix maps hazard severity on one axis and hazard likelihood on the other axis. Once a hazard's severity and likelihood are determined, they are mapped to a particular HRI matrix cell, which yields the hazard index and the relative risk for that hazard. The hazard risk level establishes who can accept the risk by authority level.

Each matrix cell contains an HRI risk index number, indicating the relative (vice absolute) safety mishap risk presented by a particular hazard. Note that the cells for a 5 x 4 matrix are labeled with a risk index of 1 through 20, where 1 represents the highest risk and 20 the lowest risk. The smaller the HRI number, the higher the safety risk presented by the hazard. The HRI indices are divided into four groups which comprise the High, Serious, Medium and Low risk levels in the Risk Level Table.

## B.3 Risk Rating Criteria

The suggested mishap severity category criteria from MIL-STD-882D are shown in Table B.1, and the suggested mishap probability category criteria are shown in Table B.2.

Table B.1 – Hazard/Mishap Severity Categories

| Description | Cat | Mishap Definition |
|---|---|---|
| Catastrophic | I | Could result in death, permanent total disability, loss exceeding $1M*, or irreversible severe environmental damage that violates law or regulation. |
| Critical | II | Could result in permanent partial disability, injuries or occupational illness that may result in hospitalization of at least three personnel, loss exceeding $200K* but less than $1M*, or reversible environmental damage causing a violation of law or regulation. |
| Marginal | III | Could result in injury or occupational illness resulting in one or more lost work days(s), loss exceeding $10K* but less than $200K*, or mitigatible environmental damage without violation of law or regulation where restoration activities can be accomplished. |
| Negligible | IV | Could result in injury or illness not resulting in a lost work day, loss exceeding $2K* but less than $10K*, or minimal environmental damage not violating law or regulation. |

Table B.2 – Hazard/Mishap Likelihood Categories

| Description | Level | Specific Individual Item | Fleet or Inventory |
|---|---|---|---|
| Frequent | A | Likely to occur often in the life of an item, with a probability of occurrence greater than $10^{-1}$ in that life. | Continuously experienced |
| Probable | B | Will occur several times in the life of an item, with a probability of occurrence less than $10^{-1}$ but greater than $10^{-2}$ in that life. | Will occur frequently |
| Occasional | C | Likely to occur sometime in the life of an item, with a probability of occurrence less than $10^{-2}$ but greater than $10^{-3}$ in that life. | Will occur several times |
| Remote | D | Unlikely but possible to occur in the life of an item, with a probability of occurrence less than $10^{-3}$ but greater than $10^{-6}$ in that life. | Unlikely, but can reasonably be expected to occur |
| Improbable | E | So unlikely, it can be assumed occurrence may not be experienced, with a probability of occurrence less than $10^{-6}$ in that life. | Unlikely to occur, but possible |

Table B.3 contains the HRI Matrix and Table B.4 contains the Risk Acceptance criteria. Note that these mishap severity and probability criteria provide guidance for a wide variety of programs. The criteria in these tables can be tailored as appropriate for a particular program. If tailoring is required, a mutual understanding of the modified terms among the stakeholders is required.

Table B.3 – HRI Matrix

| Probability | | Severity | | | |
|---|---|---|---|---|---|
| | | 1 Catastrophic | 2 Critical | 3 Marginal | 4 Negligible |
| A | Frequent | 1 | 3 | 7 | 13 |
| B | Probable | 2 | 5 | 9 | 16 |
| C | Occasional | 4 | 6 | 11 | 18 |
| D | Remote | 8 | 10 | 14 | 19 |
| E | Improbable | 12 | 15 | 17 | 20 |

Table B.4 – Risk Acceptance Matrix

| Risk Level | Index Values | Acceptance Criteria |
|---|---|---|
| High | 1 - 5 | Not allowed unless signed by CAE |
| Serious | 6 - 10 | Not allowed unless signed by PEO |
| Medium | 11 - 17 | Allowed; signed by PM |
| Low | 18 - 20 | Allowed; signed by PM |

B.4 Risk Acceptance Using the HRI Matrix

The acceptance of risk presented by identified hazards is an established process set forth in MIL-STD-882 which requires that all identified hazards be eliminated or reduced to an acceptable level of risk. In addition, the Office of the Secretary of Defense (OSD) policy clearly specifies the requirement for system safety and a formal risk acceptance process. High and Serious risk hazards can be accepted only by higher levels of authority in order to ensure that these levels of risk are not covered up and that the appropriate considerations are made in the decision. Having a person "sign off" and accept risk assures accountability and responsibility for his/her decisions. The adage of "when everyone is responsible for safety, no one is responsible or accountable" is eliminated when a person is required to sign on the bottom line that the system safety process has been followed.

# AUTHOR BIOGRAPHY

Mr. Ericson has over 45 years of experience in the field of system safety, software design, software safety and Fault Tree Analysis (FTA). He holds a BSEE from the University of Washington and an MBA from Seattle University. Currently he works for the URS Corporation (formerly EG&G Technical Services) in Dahlgren, VA. He provides technical analysis, consulting, oversight and training on system safety and software safety projects. He currently supports NAVAIR system safety on the UCAS and BAMS unmanned aircraft systems, and he is assisting in writing NAVAIR system safety policies and procedures. Prior to joining URS, Mr. Ericson worked at Applied Ordnance Technology (AOT), Inc. of Waldorf, Maryland, where he was a program manager of system and software safety. In this capacity he directed projects in system safety and software safety engineering.

Prior to joining AOT, Mr. Ericson was employed as a Senior Principal Engineer for the Boeing Company for 35 years. At Boeing he worked in the fields of system safety, reliability, software engineering and computer programming. Mr. Ericson has been involved in all aspects of system safety, including hazard analysis, FTA, software safety, safety certification, safety documentation, safety research, new business proposals and safety training. He has worked on a diversity of projects, such as the Minuteman Missile System, SRAM missile system, ALCM missile system, Morgantown People Mover system, 757/767 aircraft, B-1A bomber, AWACS system, Boeing BOECOM system, EPRI solar power system and the Apollo Technical Integration program.

Mr. Ericson has taught courses on software safety and FTA at the University of Washington. Mr. Ericson was President of the System Safety Society in 2001-2003, and served as Executive Vice President of the System Safety Society, and Co-Chairman of the 16th International System Safety Conference. He was the technical program chairman for the 1998 and 2005 International System Safety Conferences. He is the founder of the Puget Sound chapter (Seattle) of the System Safety Society. In 2000 he won the Apollo Award for safety consulting work on the International Space Station, and the Boeing Achievement Award for developing the Boeing FTA course. Mr. Ericson won the System Safety Society's Presidents Achievement Award in 1998, 1999 and 2004 for outstanding work in the system safety field.

Mr. Ericson has presented training courses in system safety, software safety and FTA in the U.S., Singapore and Australia and has presented numerous technical papers at safety conferences. Mr. Ericson has published many technical articles on system and software safety and is currently editor of the Journal of System Safety (JSS), a publication of the International System Safety Society. Mr. Ericson is author of NAVSEA Weapon System Safety Guidelines Handbook.

Mr. Ericson can be reached through his website www.risk-logic.com. Other books published by Mr. Ericson include:

1) Hazard Analysis Techniques for System Safety, July 2005, Wiley.
2) Concise Encyclopedia of System Safety: Definition of Terms and Concepts, July 2011, Wiley.

www.ingramcontent.com/pod-product-compliance
Lightning Source LLC
Chambersburg PA
CBHW051530170526
45165CB00002B/678